T0350216

A Lady Mathematician in this Strange Universe:
Memoirs

A Lady Mathematician in this Strange Universe: *Memoirs*

Yvonne Choquet-Bruhat

Emeritus Professor of the University Pierre et Marie Curie (Paris, France), member of French Academy of Sciences and American Academy of Arts and Sciences

World Scientific

NEW JERSEY • LONDON • SINGAPORE • BEIJING • SHANGHAI • HONG KONG • TAIPEI • CHENNAI • TOKYO

Published by

World Scientific Publishing Co. Pte. Ltd.

5 Toh Tuck Link, Singapore 596224

USA office: 27 Warren Street, Suite 401-402, Hackensack, NJ 07601

UK office: 57 Shelton Street, Covent Garden, London WC2H 9HE

Library of Congress Cataloging-in-Publication Data
Names: Choquet-Bruhat, Yvonne, author.
Title: A lady mathematician in this strange universe : memoirs / Yvonne Choquet-Bruhat (University Pierre et Marie Curie, France, & French Academy of Sciences, France, & American Academy of Arts and Sciences, USA).
Other titles: Mathématicienne dans cet étrange univers. English
Description: New Jersey : World Scientific, 2017.
Identifiers: LCCN 2017045460| ISBN 9789813231627 (hardcover : alk. paper) | ISBN 9813231629 (hardcover : alk. paper)
Subjects: LCSH: Choquet-Bruhat, Yvonne. | Mathematicians--France--Biography. | Physicists--France--Biography.
Classification: LCC QA29.C534 A313 2017 | DDC 510.92082--dc23
LC record available at https://lccn.loc.gov/2017045460

British Library Cataloguing-in-Publication Data
A catalogue record for this book is available from the British Library.

A Lady Mathematician in this Strange Universe
(Translated from *Une Mathématicienne dans cet étrange univers*, by Yvonne Choquet-Bruhat, Originally published by Odile Jacob in 2016: © Odile Jacob, 2016)

For any available supplementary material, please visit
http://www.worldscientific.com/worldscibooks/10.1142/10754#t=suppl

Printed in Singapore

CONTENTS

PROLOGUE

I was lucky to have intelligent and cultured parents, one literary and the other scientific, who loved their children and wished to see them benefit from their own theoretical and practical knowledge. As soon as my reason awakened, I wished to understand something of this strange universe in which we live, and what we human beings do in it, myself in particular. I still see myself in my parents' bedroom asking them a question about the particular strangeness that for me was the fact that I had direct consciousness only of myself. My mother said to my father, "She cultivates solipsism." To which my father responded, "Metaphysics makes you mad." My question and the answers still seem relevant. However I did not cultivate solipsism, a learned word that solves nothing. I always believed that there exists, outside myself, a reality which I, though barely out of childhood, wished to understand. I remember once having asked my revered father, "What is light?", a difficult question for even such a scholarly optical physicist to answer to an unenlightened youngster of about 10 years of age. I don't precisely remember his answer, though there was one. In my teens I flirted with various domains of knowledge: the natural sciences, physics, history, philosophy. The latter encompassed, in principle, all the sciences, a generality that nowadays has become practically impossible due to the enormous growth of each of them. In the end I became a mathematician working on problems posed by physics.

The mathematics I have used is a construction of the human mind, starting with definitions to build, by logical reasoning, an edifice of theorems, elucidating properties of the objects defined and the relations between them. I know that the foundations of the mathematics I use are based on unprovable propositions of set theories. Others have been constructed, non standard analysis for example, but their use has been less fruitful. I did not take part in the philosophical debate of whether, as Alain Connes says and Plato had already thought, mathematics actually exists independently of its use by human beings. This metaphysical problem does not really interest me because what, after all, does it really mean "to exist"?

The experimental physicist observes the world around him and builds more and more complicated instruments to observe it at all the scales that he can. The work of experimentalists has not ceased to reveal that what, for us, constitutes reality, is nothing but an image, at our scale, of a deeper, extraordinarily more complex reality. I remember my astonishment when, looking at a green fabric with a magnifying glass, I saw a white surface covered with yellow and blue dots. Quantum physics now tells us that our reality is in fact the sum of an infinitude of others. What for us is space, and even time, is only a construction of our senses and minds. The fundamental bricks of an ultimate reality will most probably remain inaccessible to human beings, at least as long as they are actually alive.

The theoretical physicist proposes a correspondence between the phenomena observed by experimentalists and mathematical objects, sometimes in a not completely rigorous way. The mathematical physicist, who is often also a theoretical one, tries however to give a precise mathematical meaning to these objects and to the laws that their evolution obey. He appeals to existing theorems or proves new ones to establish the mathematical properties of the theoretical objects considered and seeks to put them into correspondence with the prior observations of experimentalists. The mathematical physicist also predicts new properties of the phenomena under study, deduced from those of the mathematical objects which represent them. These predictions will then eventually be observed by experimentalists, with the help of more and more sophisticated apparatus. This poses great satisfaction for the mind. The results obtained through experiment often have practical uses as well.

Among theoretical physicists, one could distinguish, as Gustave Choquet did for mathematicians, strategists and tacticians. Strategists propose a new theory — general relativity, for example, or the existence of quarks — anticipating its verification by possible observations. Tacticians, on the other hand, critically analyze the theories proposed by strategists, deducing their consequences and foreseeing their possible experimental verification with instruments one has or might build. In physics, I am essentially a tactician, using the Einstein equations of General Relativity to prove the existence of the gravitational waves that were only recently observed after about fifty years of experimental efforts.

As experienced by all researchers, a discovery, even small, is always a great joy, like that of finding an unknown land. I always found my own work on the local and global properties of the solutions of the Einstein equations — either in vacuum or coupled to good relativistic equations for sources — to be very exciting. In this last regard I had to become something of a strategist which was quite interesting as well. I was mainly motivated by the wish to better understand a particular aspect of our reality. I must confess, however, that I occasionally let myself be tempted, out of sheer intellectual curiosity, by some works in pure mathematics: to solve a clearly posed challenging problem, even one without any apparent contact with physical reality, is always a pleasure.

In this autobiography, written on the occasion of my retirement from scientific work in my nineties, I mix evocations of scientific ventures with stories of the near and distant travels they have occasioned, and of the scientists in all those countries I came to know. I intersperse this account with that of events from my private life. The essential part of my life has been my three children, whose births range from 1950 to 1966. At their request, I have written these pages.

1

ANCESTORS

The Bruhat

My father's ancestors, I have heard, lived in Auvergne, a province in the center of France where the Gaul army, led by Vercingetorix, had been defeated by Jules Cesar in forty-six before Christ. My great-grandparents lived near the city of Brioude, which enjoys a beautiful Roman cathedral. My great-grandfather had two children, Antoine, my grandfather, and his younger sister, Marie. My sister, my brother and myself called her Aunt Marie. According to Aunt Marie, my great-grandfather was the twin of the twelfth and last child of his parents. Growing up, he worked by breaking stones for roads. As a child I was impressed by the social upheaval of the Bruhat family from the profession of stone breaker on roads to a comparatively higher one of colonel and school teacher. Later, my brother told me that our ancestor had, in fact, owned a small movable firm. He had married a girl named Pizel, of whom I know nothing. Aunt Marie had kept in contact with some of Pizel's cousins. One of them emigrated in Algeria as a school teacher. He had taken his job very seriously, and, according to Marie, had done his best to help his students, as she did for her students in the school of the school of the French Republic in Congy, a village in Champagne. My father, already an adult, came to know another physicist named Bruhat coming from Brioude. They discovered they were cousins and became friends.

My paternal grandfather, Antoine Bruhat, had made a career as a military man, rising from the ranks. He was retired when I knew him. He still had great bearing, tall for the times, and he kept his bearing quite straight. He had white hair and a beautiful white mustache. We often referred to him as a colonel. I learned later that he was only a lieutenant-colonel, because he did not have the diploma required to be an officer. While he was a young soldier garrisoned in Besançon, he had married Jeanne Aberjou, a girl of the local upper middle class. He had made to her nine children, all still-born except the second, my father, who was born quite small after only seven months of pregnancy. The others, born on time with a normal weight, could not get out from the maternal womb without being cut up. This story, told by Aunt Marie, had horrified my mother. She did not forgive her father-in-law for making nine children to his wife under these conditions. This unfortunate woman died at about forty-five, at the beginning of the First World War, probably from uterus cancer. She had only the assistance of her sister-in-law, Marie, as her husband and son were in the army.

My grandfather did not speak much with his grandchildren, and my father did not like to evoke painful memories. I know nothing of my grandmother's family. I think it was socially higher than my maternal grandparents, Hubert, because my mother told me she had asked her father, a school teacher, not to show nor feel any inferiority when meeting the colonel Antoine Bruhat before her marriage with my father. She regretted that he had felt so before the marriage of my maternal uncle, René, with the daughter of a professor in the Sorbonne. The encounter between Eugène Hubert and Antoine Bruhat went smoothly, though Eugène, as a dedicated teacher of the public school, was Dreyfusard, while Antoine was anti-Dreyfusard, as a faithful member of the army. Both were civilized men and satisfied with the marriage of their children.

Aunt Marie told us that Georges, my father, was a boy devoted to his mother. He spent many hours quietly, near her, when she had to stay in bed because of the repeated tragic deliveries. He did not speak to us about his mother, but gave her first name, Jeanne, to his eldest daughter. Marie liked her sister-in-law, a woman who married young, and possessed culture and a sweet temper. However, she did not feel very close to her as their characters and lives were too different.

The Hubert

What I know of my roots mainly come from my mother's ancestors. They were workers or farmers in villages in the vicinity of Mantes-la-Jolie, a city of île de France. Among my great-great-grandparents, only one had been a foreigner. He was named Galeppi and came from Lugano, in the Italian part of Switzerland. A worker making his traditional "tour de France", he had fallen in love with a great-great-grandmother of mine and married her. He settled in France, near his wife's family, in Dammartin en Serve, a village 3 km from Longnes. His birthplace, which seemed very far away in those times, and his merry character had kept alive his memory among his descendants. His daughter, my great-grandmother, married a M. Lainé, of whom I know nothing. They had two children: my grandmother Louise, and a younger daughter named Berthe who died from diphtheria at eight. My mother was named after her. When widowed, Mme Lainé came to live with her daughter. My mother, her granddaughter, was very fond of Mme Lainé. She described Mme Lainé as a very open and warm-hearted woman, probably like her Swiss-Italian father. Fairly or otherwise, my mother blamed herself for her grandmother's death. The old lady had come to see her in Sévres, and my mother had taken her for an outing in a cab. She died from pneumonia a few weeks later. My mother believed it had been the result of a cold caught then.

The ancestors of my grandfather Hubert owned a small farm at La Fortelle, a hamlet near the village Longnes, fifteen kilometers from Mantes. They were hard working small farmers. I don't know much about them, though Longnes is only some sixty kilometers from Paris. I know that my great-grandfather Hubert had at least another son apart from my grandfather; for my mother he was "Uncle Jules." He had died from diabetes, as did later, his junior, my grandfather Eugène. Some years before his death Uncle Jules had been amputated, first from one leg, then from the other, because of gangrene. He had valiantly coped with these amputations.

I have not had the opportunity to know my grandfather, Eugène Hubert, who died when I was still a baby. He was, according to my mother, a remarkable man of great intelligence and humanity. Born and brought up in a family of small farmers, his exceptional qualities were recognized by his school teacher who obtained, for him when he was fourteen, a position called "student-master". There was, in those times, one school in each village with one class for students of all grades. The "student master" was in charge of the cleaning up of the classroom and the lighting and functioning of the stove. He was also in charge of part of the teaching for younger students. In exchange for these duties, the "student master" had a small indemnity for food and perhaps, lodging. He could prepare for the examination to become a school teacher, which my grandfather promptly did and got the diploma. A few years later he was promoted and posted to Paris. He then went to live near the canal Saint Martin, with my grandmother. He was appreciated by his students and colleagues. They gave him the responsibility of managing the funds to help persons with difficulties, in particular with sickness. My mother thought that her father might have caught tuberculosis from their files; he died from it when he was only sixty years old.

This former "student master" fathered two children. A boy, René, born in 1885, and seven years later, with the prompting of his spouse, my mother, Berthe. Both were admitted to study at "École Normale Supérieure".

My grandmother, born Louise Lainé in 1860, had left school early to work in the esparto factory of Dammartin en Serve, which made carpets and other objects with rope. A museum has recently been created in its former place, exhibiting memories from these times. My grandmother did not even have the "Certificat d'Etudes". She had, however, good handwriting and a knack for not making spelling mistakes. She stopped working outside of home after her marriage with my grandfather. I know nothing of the circumstances of this marriage, a union between a young workwoman and a school teacher from a nearby village. I think that Louise Lainé had been quite a pretty girl. In her old age she had kept very blue eyes and white and fine skin in spite of its wrinkles. My grandmother did not speak much about herself. I remember she confided to me that she had been sad to remain the only child of her parents after the death of her little sister. That was why she gave my mother the first name of that lost sister. My grandmother never confided in me about her married life. I know only that she had once told my mother, who repeated it to me, "I have been happy two years," without further explanation. I remember she told me she had strongly insisted for her husband to give her a second child, my mother, and the pleasure she had had feeding her at the breast, a thing she had not been able to do for her son.

My mother's family lived together in Paris after Eugène was appointed there. They spent their holidays in the small house at Dammartin en Serve that my grandmother had inherited from her family. She remained, always, very fond of that house. The house, enlarged off the neighboring one, has belonged to my parents, then to their children who held family meetings there. It belongs now to one of their granddaughters who lives near Paris.

2

GOOD DADDY, AUNT MARIE, MAME AND TONTON

Good Daddy

When I knew my grandfather Antoine Bruhat, he lived in a comfortable rented apartment on rue d'Artois in Paris, near "La Madeleine". We called him "Bon-papa", that is, "good daddy". My mother, who did not much like her father-in-law — without showing it in the presence of her husband — once asked us ironically: "Is your father bad daddy?" Bon-papa had a governess of a certain age who also lived in the apartment. She called my father "Monsieur Georges". Her name was Suzanne and she was in charge of cleaning and cooking. My father was not very close in spirit to his own father, but he was his only child, and very conscientious. We saw Bon-papa regularly. We sometimes went to have lunch at his place; he kept some toys for his grandchildren. I remember a toy grocery store which Suzanne furnished with real products. I remember that once, Bon-papa took us, my sister and I, to the officers' club to see some performance that I have long forgotten, but I remember vividly how proud I was to have been taken there. My grandfather came fairly often to lunch at our place, or to spend some time with us at Dammartin or some other place where we holidayed. He died in 1935, soon after the nomination of his son at the scientific direction of the École Normale Supérieure, which had made him very proud. He was only eighty years old. He used to say, "My father lived until he was ninety-two, and he did not take care of himself; I do!" He had invested a fairly important sum in an annuity, a short time before his death. According to my mother, he had started acting imprudently after getting acquainted with a woman younger than him. Suspicion of the existence of this late-in-life friend came to my mother from the sight of an unknown lady who put, discreetly, a bunch of violets on the grave of my grandfather on the day of his burial. The fact is that he had a stroke. I had been shocked when I saw Bon-papa in his apartment after this accident. He had become an old man who seldom left his armchair. The following winter, he caught pneumonia, and antibiotics had not existed yet. He died in a few weeks, in spite of the care of my mother, his sister Marie and a nun acting as a nurse. My mother had started friendly relations with the young woman, and suggested they continue after my grandfather's death, but the young nun said it was not compatible with the rules of her order.

I feared the shame of not shedding tears at Bon-papa's burial, but thanks to the environment and the music, I cried copiously.

Aunt Marie

Aunt Marie, of some years younger than her brother Antoine, had remained unmarried. She once told my mother that a young man had come to ask permission from her father to marry her when she had still been living with her parents. He had been rebuked and Marie had heard of it much later, like the Marie of the book by Germaine Acremant "Ces Dames aux Chapeaux

Verts". In her case, there had been no Arlette to straighten out destiny. Marie remained with her parents and took care of them in their old age, until their death. She told this story to my mother without acrimony.

Aunt Marie lived in Congy, a small village in the neighborhood of the city of Epernay in Champagne, where she had been a school teacher until her retirement. She knew everybody there; many had been her students and remained fond of her, but we were her only family, as my father was her only nephew. Marie had a happy temperament. She never complained. She would laugh at her own jokes. She liked to speak of her daily life and reminisce. My mother, a woman hard to please, liked her. Marie came to stay with us every year for a few weeks, either in Paris or in Dammartin. It is from Aunt Marie that we heard of the Bruhat family and my father's childhood.

My mother had a grudge against the whole world after the death of my father, and she stopped seeing most of her friends. However, Aunt Marie's visits, unchanged, remained as welcome as before. She stayed with us for a month each winter until her advanced years forced her to leave — first her house, then the retirement home where she had kept some autonomy. She had chosen that home close enough to Congy to receive visits from the friends she had made there — that is, those who were still alive, who became few, even among her former students. Before making the choice of this shelter for her old days, Marie had come to Paris and visited several retirement homes with my mother. One of them, fairly close to our apartment, seemed convenient to her, but she finally chose a retirement home near Congy. She told my mother, "Here I will know only you." My mother was a bit hurt, though she understood and was perhaps relieved.

Marie kept a bed, wardrobe, table and armchair to furnish the bedroom of her new residence. She gave us the remainder for

Dammartin, where furniture was missing since the fate (firewood) given by the German army to that inherited from Bon-papa. We let Aunt Marie believe that the furniture of her living-room was going to furnish the room of the same name in Dammartin. She was very pleased with that. In fact, it furnished, very nicely, the large kitchen where we had our meals when there were not too many of us.

Marie was satisfied with her new home. Unfortunately, time continues to run. Marie had to leave for the geriatric hospital of Epernay after she broke a bone. Her friends, dead or too old, did not visit her anymore. Her grandnieces and grandnephew, who lived too far and were too busy, did not come often. In agreement with her character, she did not complain but asked that we write to her. After her death, I found in her drawer, not without remorse, the too-few postcards I had sent to her. I remember, with gratitude, my visit to Aunt Marie after my divorce. A practicing Catholic, she could have blamed me. She did not, but instead gave me her best wishes of happiness. The furniture of her room in the retirement home are now in my house, the only ones coming from my youth.

Aunt Marie lived to a hundred and three. I remember the ceremony organized to celebrate her centenary at the geriatric hospital where she had shared a room with other old ladies and had moved by pushing a chair. Aunt Marie had kept her head. Several times she interrupted the speech of the mayor celebrating her anniversary by saying in a kind tone, "No, mister mayor, it was not quite like that." That day, one of her roommates told me, "She raises the spirits of us all." After the ceremony, however, when I wished her many more years to live, she said, "You wish me much harm." She died peacefully, not awakening from the coma she had entered in her sleep.

Mamé

My grandmother, whom we called Mamé, lived with us since she was widowed in 1925 until her death in 1952, at ninety-two years of age. She was a discreet and undemanding woman. She was on good terms with her son-in-law, rendering him small services — for instance, re-sewing his buttons. They never said anything unpleasant to each other.

When we lived in rue d'Ouessant, Mamé accompanied Jeanne and I back and forth to our school, the Cours Gernez on avenue de Suffren. On Thursdays, when we had the school day off, she would take us to the nearby Champ de Mars and watch us play. After we moved to rue d'Ulm, we did not need her to accompany us to move around. My grandmother decided not to go out of her new lodgings. Before she had a cataract, which was badly treated at the time, Mamé made herself useful by assuming various small duties. For instance, each Monday she made sure that Jeanne and I had the required clean pinafores with our names embroidered in red cotton, for school. To pass the time she avidly read her favorite weekly paper "Les Veillées des Chaumières" or novels we were lending her. She did not confide much in us about her past, nor did she give advice to my mother on our education. However, she was a very sensible woman. She believed in God without practicing a religion. She made, to her grandchildren, critical comments on their behavior: finish your bread, put your hands on the table and so on. She had some key historically-inspired sentences which I like to quote to my children whenever appropriate. One is "And why did we fight for three days?", a reference to the three days during the revolution of 1830 which overthrew the king Charles X and brought to France, in principle, freedom in action for all. Another, commonly said after enjoying a good meal, is "At least, Prussians will not take that", referring to the 1870 war. My grandmother was ten at that time but, though her family had been of small means, the leftist movement at the end of that war — or "la Commune" in France — was recalled to be a terrible period where bandits fought against the legitimate government.

My mother assumed her duty as a daughter, as was usual then, but she did not feel very close to her mother. Their personalities were too different.

Tonton and Tantine

My father was an only child. My mother had one brother, he was seven years older than her but they were very close. His nieces and nephew called him Tonton; Tantine was his spouse when we knew him. They were an important part of my youth. My sister, my brother, and I spent time with them in Poitiers during the school year of the Phony War: 1939–1940. After the war, I stayed at their house in Strasbourg several times while I was working on my thesis under the direction of Lichnerowicz, a professor there. This is why I include Tonton and Tantine in this chapter.

My uncle, René Hubert, the cherished elder brother of my mother, was born in 1885. He performed brilliantly in his studies at the lycée Rollin, and was crowned with six nominations at general competitions between 1900 and 1903. He entered the letters section of the "École Normale Supérieure", in 1905. He came first in the Agrégation of philosophy in 1908. As was the rule at that time, he went to teach in a lycée; first in Chambery, then in Perigueux, and finally in Marseille in 1912.

As with all men of his age, René was in the army during the First World War: first in the infantry, then as the commandant of a company of machine-guns. He was wounded on 1918, December 5. As with many men of his generation, the war in the trenches was a terrible ordeal. It led René to deep thoughts and philosophical writings on wars and human nature. In 1919, René Hubert got a position in Lille; he was promoted to the position of a full philosophy professor after defending his thesis in 1923. He was elected Dean of the University of Lille in 1931 and 1934 before being appointed as Head (in French, we say "Recteur") of the University of Poitiers in 1937.

My uncle had married Germaine Rodier, the very young daughter of his professor in the Sorbonne. My mother had no sympathy for the girl, given that she was brought up by rich parents and had no profession. I don't know if it was the same for all the professors in the Sorbonne at the beginning of the twentieth century, but the Rodiers lived in style. They had two full-time servants, a four-wheeled carriage and a coachman to drive it. The young couple quickly had a son, Iannis. My mother, still very young, enjoyed the smiles of the baby. But the relationship between the couple soon turned sour, and a few years later they divorced. I have known Iannis only as an adult. I was moved by his likeness to his father.

In my childhood, my uncle was an important person for me. He was divorced and lived alone in Paris, at rue du pont de Lodi. He came often to see us at rue d'Ouessant. My mother loved her elder brother, and he returned her affection. He was a man of fine presence, and rather tall. He had an open mind that was very cultivated and curious of everything. His thesis was on the philosophy of history and the issue of social origins. He was a specialist of the Encyclopédistes, particularly Jean-Jacques Rousseau, whose independent mind agreed with his. My uncle was also a man full of humor. He loved his nieces who worshipped him. He called us "l'araignée trublionante", something like whirling and trouble-making spider (my sister), and "le cloporte inassouvi", the unsatiated sow bug (me). He never failed to give us a beautiful book for Christmas. He participated in all family celebrations. We called him Tonton. When I learned he was marrying again, with a Corsican seventeen years younger than him named Georgette Bonardi, I asked him if we should call her "Tontonette". He answered, "No, Tantine." I got along well with Tantine. She taught natural sciences before her marriage. She was knowledgeable in botanical sciences. I remember, with gratitude, strolls in the neighborhoods of Dammartin where she told me the names of flowers and explained their classifications and botanical properties. Still now I see myself again, with clarity and pleasure, in a meadow or a wood at Dammartin with Tantine showing me an orchid, an orobanche (broom-rape), an euphorbe (spurge), a colchique (colchicum) or some other flower.

In 1936, my uncle was appointed Recteur of the University of Poitiers and left Paris, but he came to stay with us when his professional duties called him, fairly frequently, to Paris. He came also to spend a few days of holidays in Dammartin with his young wife and their little girl, Madeleine, whom we called Mado. He was happy to be, for some time, with his mother. She was also happy to see her son, and enjoyed his affectionate teasing. He made us laugh by calling her by her first name, Louise. Mado was a nice and charming girl, closer to my brother François than to me: she was only a few years younger than him, but he himself was my junior by five years. Her parents doted on Mado. After marrying, Tantine stopped teaching and devoted all her life to her daughter. After the death of her mother, Mado told me she felt very lonely. She said, "We were so close to each other." Tantine dutifully fulfilled her obligations as housewife and spouse of Recteur.

The moral and political convictions of René Hubert could not agree with the Vichy government's ideals. He was forced into retirement soon after the armistice. I think it was lucky for him. He thus escaped jail and the deportation camp which would perhaps have been his fate otherwise, as it happened for my father and other persons with responsibilities who did not obey the orders of the occupying German authorities and did all they could to protect others. My uncle used his forced leisure to return to the pursuit and writing of his philosophical works, in particular, on education. His book on this subject, completed by Miallaret, has been republished several times.

After France's liberation, my uncle returned to his position in Poitiers, but was soon appointed Recteur of the University of Strasbourg in 1946, an important position deserving of an opponent to the Vichy regime. The professors in Strasbourg were Germans during the five years of the occupation. The French professors who were there in 1940 had been transferred to Clermont-Ferrand; they came back to Strasbourg after its liberation.

My uncle continued to come to stay with us when he came to Paris. It was a ray of light in my painful life at that time. I believe it is due, at least partly, to him, that the Minister of Education found for my mother a position, when she had to leave our apartment on rue d'Ulm. The newly-appointed scientific director, the chemist, Dupont, was in a hurry to occupy it. My uncle also obtained the red ribbon of the "Legion d'Honneur" for my mother. She certainly deserved this decoration, but it was given to her to compensate the rejection of her request to award a higher decoration posthumously to my father. We felt it bitterly. My father had done much more to resist the enemy and help his endangered fellow citizens to survive than some colleagues who had left for England, and it had caused his death! However, my mother was pleased by that red ribbon. She was proud of it. I was driving my recently acquired car, a "deux-chevaux", with my mother sitting next to me, when some breach of the highway code led the police to stop me. When I was trying to negotiate matters with the policeman, my mother got out of the car to support me, and the policeman let us go without a charge. My mother said, "He saw that I have the Légion d'Honneur." This incident is a happy memory.

My aunt had, of course, followed her husband. But this Corsican did not much like Strasbourg, a city with very hard winters and inhabitants who talked in a dialect quite different from the one spoken by her ancestors. I remember Tantine had made us laugh by telling us that one day, entering a shop, she had asked to see the boss. She had understood the answer of the assistant to be, "He is dead, he comes only sometimes", and it had rather frightened her. Myself, I was happy to stay at my uncle's house when I went to Strasbourg to talk about the progress of my thesis with Lichnerowicz. Strasbourg is a beautiful city; the riversides there have a particular charm. I appreciated them again when, in the beginning of the seventies, I was on the way to the seminars on mathematics for physics organized by Paul Kree, my colleague in Paris. Jean Leray, Paul Kree and I often travelled together; time passed quickly. I remember a train return trip when, starving, the three of us shared a delicious kugloff, a specialty of Strasbourg which Kree intended to bring back for his wife.

During these seminars I could not stay with my uncle because, unfortunately for him and those who loved him, Tonton had been struck a short time before his retirement by a heart disease. He died from it in 1954. Then, I was still a professor in Marseille. Tantine took care of her husband devotedly. After his death she came to live in Paris with Mado, who was preparing, in the Sorbonne, a thesis on Greek literature. Mado admired the ancient Greece civilization, like her father, who probably had influenced her choice. As the philosopher he was, however, my uncle had been somewhat disappointed by the overly grammatical orientation recommended by the thesis adviser. My departure for the United States, and the illness and death of my uncle have prevented me from following the details of the work of my cousin, but Mado became Docteur ès Lettres, and taught in the Sorbonne. She married a former student of the famous school Polytechnique, whom she met at the club of bridge she attended, accompanied by her cousin François. I think she had had a little soft spot for François, but he had for her only brotherly feelings.

3

MY PARENTS

My Parents

My childhood was happy. I was loved by parents who never quarreled and wished the best for their three children. However, nothing is perfect on this Earth, including human beings.

My parents had met soon after the First World War, in Lille where they were both professors: my father, in the physics department of the university, and my mother, in the senior year of the high school, teaching literature. They got to know each other thanks to my uncle René, who had known my father at the École Normale. At that time, the small number of students allowed for cordial relations between classes of nearby years and of different subjects. Through René, my mother-to-be, Berthe, knew my father-to-be, Georges. She told me that after several chaste outings with him, bad at that time for the reputation of a girl, she told Georges she could not continue to go out with him; he answered at once, "Of course, we must get married." I think they were both ready to start a family. The marriage took place soon after and they went to Switzerland for their honeymoon, as organized by Georges. Berthe told me (much later) that she would have preferred the seaside, but had said nothing of it to her fiancé.

One can say they were a happy couple, sharing the same bed, respecting each other and wishing the best for their spouse. Our family life was, on the whole, happy. I never heard my parents quarrel. Lunches and dinners were taken together: the three children, their parents and their grandmother sitting at the same table in their usual places. I was sitting near Mamé, facing the light, a privileged place given to me to, supposedly, compensate my visual problem. Mamé did not speak much, but the others discussed and told of the events of their day. I still have some nostalgia regarding these friendly moments. My brother François did not speak much, but Jeanne, my sister, was prolix. My father complained sometimes, but without screaming, that he could not get a word in. The only mar to these happy memories were the sometimes violent quarrels between Jeanne and our mother. I remember one day when, leaving the dinner table after a heated quarrel in which our mother was not without fault, dad said to Jeanne to bring peace back, "You know your mother, calm down."

My father was a gentle and reasonable man. I never heard him complain about his wife. My mother appreciated the intelligence of her husband, but she was sorry that he did not have more time for other activities than his work, nor more interest for literature. My mother was a literator, she had a subscription in a lending library and read a lot, mostly contemporary novels written by great writers, because she already knew the others. My father had little time to read, and his character was not really given to the brilliance in conversation shared by my uncle René and specialists in literature like the writer Jean Guehenno, husband of a fellow student of my mother, or some mathematicians like Henri Villat or Gaston Julia. They visited us sometimes and my mother liked to talk with them. She was very disappointed by the pro-Nazi behavior of G. Julia during the war and his refusal to attempt an intervention to save my father from the claws of the gestapo.

My mother had a strong personality, but was very feminine. Though an independent spirit, she did not aspire towards independence. She told me, several times, that she reproached my father for him failing to make her feel protected. However, she was. My father was not a man to impose his authority, but he participated actively in decisions concerning the health or the studies of their children. He accompanied his wife to the doctor or the dentist. My father managed the budget and invested the savings. He took care of the small repairs in the apartment.

Household problems, daily budget and menus were my mother's domain. The cooking on the days that the maid was absent was solely her responsibility. My mother told me that at the beginning of her marriage, she had put a dish to cook on too strong a fire before leaving the kitchen. Sometime later, my father came to tell her, "There is smoke in the kitchen." My mother asked, "You have lowered the fire?" My father answered, "No, I have opened the window."

The summer holidays were long, from July Fourteenth to October First for us children and for my mother, a professor. They were shorter for my father, who was not only a professor but also a researcher and director of a great laboratory. We spent the Easter vacations and parts of July and September in our great house in Dammartin, my father joining us sometimes, only for the week-ends. My parents decided, together, where the family would spend the month of August, far away from Paris and Dammartin. Mamé would stay in her house in Dammartin with the maid, who would take her vacations before or after August. My mother preferred the seaside, my father, the mountains. In principle, we went alternately to one or the other. My mother told me, years after the death of her husband, that it had been twice as often we went to the mountains. I have not verified this. I don't remember quarrels on this subject. I know it was my father who was in charge of renting houses or making reservations in hotels where we would stay for those holidays. He also took care of the booking of taxis and excursions. A wonderful early memory is the crossing of the lake of Come by boat and the view on the Boromées islands which revealed to me the joy that the beauty of nature can give. I remember well the holidays of my adolescence, in the last years before the war. All have left me happy memories, regardless of whether they were at the mountains or seaside. I remember holidays in Grindelwald, at the foot of the impressive mountain, the Cervin. I remember in particular climbing amidst a splendid landscape with my father, and my sister Jeanne (who, in contrast, did not much like sports), all of us roped together with a guide. François was too young to come with us. My mother could have joined us with him at the summit by taking a cable car. But she had preferred, not without complaint, to stay at the bottom to avoid, according to her, exposing her dear little one to this danger! Our last family holidays at the seaside were spent in Saint Raphael in 1938, with beautiful excursions along the Mediterranean coast, in particular, to Agay, in a rented cab. Our mountain holidays, again in Grindelwald in 1939, had been shortened by the war.

I will now speak a little more, separately, of my father and mother. Both have played an important role in my life.

My Father

My father, Georges Bruhat, was born in Besançon, pronounced Bzançon, "old Spanish city"'as Victor Hugo wrote. My father spoke little of his parents, the paternal authority and especially, the sufferings of his mother, being painful memories. My father preferred to tell us, sometimes at the family dinner table, some silly things he had done as a boy, like the time he set the curtains of the dining room on fire. He also told us some facts about his life at school. I remember that he wrote dissertations in Latin and that his professors gave their lessons in frock coats.

After preparing at the lycée Saint Louis, my father had been admitted, the first by merit on the list, both to the École Normale Supérieure (ENS) and to the École Polytechnique. It was as exceptionally rare a performance in his time as it is today. He had chosen ENS because scientific research was his vocation. After getting the diploma "Agrégation", he taught for a year at the lycée Buffon. At that time, the CNRS (National Research Center) had not existed yet. Then he was appointed as "Agrégé Préparateur" at the ENS where he wrote a thesis on optics under the direction of Aimé Coton. He fought in the First World War on the front lines, in a section dealing with sound detection. The improvements made by Georges Bruhat to this detection earned him the "Croix de guerre" distinction and the lifelong esteem and friendship of his chief in the army, the famous mathematician, Emile Borel. After the war, my father was a professor at the university of Lille, then at the Sorbonne in 1927. He was appointed Vice-Director of the ENS in 1935. He worked at the building of a modern, up–to–date physics laboratory. My father was very much attached to "l'École" (the ENS). He did not have much time to read, but he did not fail to follow the new adventures of the former "normaliens" (students of the ENS) Jallez et Jerphanion in the books of Jules Romains' "Les Hommes De Bonne Volonté". My father was devoted to his teaching and to his students until his death in a deportation camp on January first, 1945.

Georges Bruhat was a physicist and author of a set of remarkable books which have been republished many times. Each book has been updated by one or several colleagues who were specialists in the subject. In particular, the book on optics, the domain of the original works of Georges Bruhat, has been updated several times by the Nobel prize winner Alfred Kastler, whose works have contributed to the discovery of the laser. This book has been recently reprinted once more, with an index and a bibliography completed par P. Bouchareine. It has more than 1000 pages and its title is still "G. Bruhat Cours de Physique Générale, Optique", and in smaller print, "sixth edition by A. Kastler."

My father was a true physicist, convinced that there existed a reality that only observations would allow us to know. He was also a very intelligent man and had a logical mind, quite able to manipulate mathematical equations, in no way hostile to their use. When I was working to be admitted in the "École Normale" for girls — indeed the rue d'Ulm ENS was not, in those times, open to girls — my father liked to look at my mathematics problems, quite ready to help me if I expressed desire for it. On the other hand, if I asked a question about physics, he would often answer with a reference to one of his books. Thus, my father was not one of the physicists hostile to mathematics. Those can be forgiven because at that period, the Bourbaki school of mathematics showed deep contempt for physics. I heard Dieudonné, a great mathematician and very active member of Bourbaki, say, "If I knew that the mathematics I do are useful for physics, I would stop doing them." However, Dieudonné was a man with powerful intelligence and great honesty. I have always had good relations with him in spite of some occasional arguments in evaluation committees. My father was an example of a physicist who did not hesitate to use the mathematics useful in describing physical phenomena. His mind was open to new discoveries, having no preconceived belief in the reality that experiments strive to unveil. The works of Planck had convinced him of the reality of quanta. The experiments of Faraday and those of Michelson and Morley, which confirm the laws of electromagnetism and the constancy of the speed of light, and the observations with the telescope at Mount Wilson of the shift of spectral lines near the sun, had made my father accept the Einstein theories; first of Special Relativity, then of General Relativity, which abandon Newtonian ideas of absolute space and time. My father had understood that there are constructions of the human mind to describe reality, deduced from observations made at our scale. However, my father found the speculations uninteresting without experimental support, especially when they had mathematical formulation that was not rigorous. In the forties, the head of theoretical physics in France was Louis de Broglie. He had been a genius, but most of his students lost themselves in poorly founded speculations. My father did not hide his opinion from me. I remembered it when the time came to choose an orientation for the researcher I wished to become.

In everyday life, the man who was my father was at the image of the physicist. He was an intelligent and reasonable man, easy to live with, but well-organized and attached to his values. He loved his wife and children and found time for them, despite the time he devoted to his work. Sunday was sacred, reserved for the family. None of us would have thought to make a solo plan for that day. On weekdays, my father spent the morning in his laboratory, came back to the apartment to have lunch with us, had a short nap, and then usually went back to his laboratory until the time came for dinner with the family. After his customary cup of tea, he retired to his office to work on the writing or updating of his books in general, until about eleven o'clock. However, he did not refuse to modify this ritual if some member of his family needed him. After dinner, he did not begrudge his help with some of his children's homework. He even helped my sister with her translations from Latin when she was already a university student. On my part, I used to ask him to make me recite history or geography on the program of an examination. I never liked humdrum work; I did not learn my lessons one by one, but the whole set at the time of the examination. The recitation I made to my father of the history book by Mallet and Isaac contributed much to the good marks I got. When one of his children had a sore throat, my father lent them a "magic scarf" in silk reserved for that occasion. It was supposed to help cure it, and perhaps it did!

I remember also, that it was my father who accompanied me to the orthoptist who was trying to straighten my eyes. Indeed, when I was about two years old, I began to squint: this is called strabismus. My mother said it was due to a fall in my early childhood when, already impatient, I fell often. She had baptized me "La petite Papofski", a name inspired by the illustrations of the book "Le Général Dourakine" by the Comtesse de Ségur, where one sees the youngest of the Papofskis always flat on the ground. Of course my falls had nothing to do with my strabismus — it is genetic. An oculist said later to my father that, given his eyesight, he should have squinted. My mother took my flaw with philosophy, but my father was conscious of the handicap it would be for me later and did his best to try to cure me from it. I was operated on two times in my childhood, without much success in spite of the exercises done, watched by my father, by looking in an apparatus supposed to help obtain binocular vision. I don't recall my squint ever interfering with my interactions with people when I was a child. My friend Adeline even told me that she never noticed it. It was different later.

My father was an affectionate and devoted man; I was very attached to him. However we did not tell confidences to each other. I think that he did not like the complications they could have involved, as he did not like metaphysics. When he had some worry that concerned one of us, he would prefer to tell it to my mother who then conveyed it with a philosopher's point of view. For example, when I was preparing for entry to the ENS: "Yvonne does not work enough." Conveyed by my mother: "Your father thinks that you don't work enough." Me: "I cannot work more." Comment of my mother: "You know best."

My Mother

My mother was a literary who had achieved brilliant studies in public schools. She had had some very good professors in the lycée Jules Ferry. In particular, she had enthusiastically admired her professor of literature in "Première Supérieure" (equivalent to the first year of university), Georges Dalmeyda. This remarkable man (who was later elected Professor in the prestigious Collège de France) had welcomed my mother in his family, which was of upper middle class Israelite; a higher social level than her own family. She had learned a lot from him and felt for him and his wife a great gratitude. She had made friends with the elder of their two daughters, Madeleine. I have not known Georges Dalmeyda, as he was already dead when I was only a child, but later I often met Mme Dalmeyda and Madeleine, as my mother had kept her friendship with them.

My mother had also made friends among her classmates. In particular, Madeleine Kont, who became a Professor of History, and her colleague in the lycée Fénelon in Paris. The whole family went regularly to spend Sunday afternoon with the Moliniers. They had a daughter and a son of about my age, Jean-Pierre, a very turbulent but friendly boy who was a sports fanatic. He continued for a long time to practice sports while teaching English at a lycée in Annecy, a pretty city at the foot of the Alps at the edge of a large lake which reflects the mountains. At the Moliniers', we met another friend of my mother's, Mme Surugue, and her son Paul, a great friend of Jean-Pierre. Paul became an ophthalmologist and operated on my eyes many years later.

My mother was admitted first of her year in Sèvres, École Normale Supérieure for girls, first also at the "Agrégation des lettres" which contained literature, philosophy and English then. She became a professor of lycée (high school) in literature, at the beginning in Chambéry. She told me of her departure, at 22 years old, alone and frightened, for that unwelcoming city at the foot of the mountains, icy in winter. She soon obtained a change for Poitiers, which she liked better in spite of the presence of some adults in the classroom, convalescent Serbian soldiers (it was in 1914, the First World War). They appeared to her to be ignorant savages, though polite and respectful. She made a close friend in Poitiers, who was a few years older than herself, Léonie Estrade, a professor of arts and a painter. She had no family and became an older sister to my mother. Us three Bruhat children, born a few years later, called her "Marraine" (godmother), though she was only the godmother of the first born. Mamé had given up this place to Marraine, which, by tradition, should have been hers. Marraine came each year to spend part of the holidays with us. She was a member of the family.

After the First World War, my mother was appointed as a professor of literature in "Première Supérieure" in Lille. There, she joined her brother René, a philosopher and professor at the university where the physicist Georges Bruhat was also a professor. I already told you how this circumstance led to the marriage of my parents. I don't think it was the result of a great passion, but rather, that of much affection, esteem and a perfect fidelity. I am sure of that, at least for my mother, because she "lived without mask", as her colleague and friend Nelly Fouillet (in literature Claire Sainte Soline), said, a comment that my mother told me with pride. My mother followed her husband to Paris when he was elected in the Sorbonne. She obtained a position in the lycée Fénelon in the heart of the Latin quarter. She taught philosophy in the last year of high school, the positions in "Première Supérieure" being already taken. My mother never made a bitter comment about that. However, she often complained of being tired from her teaching. It is true that she did not spare herself when giving her lectures on philosophy. They were much appreciated by her students. Some of them still came to see her years later. One of them, Renée Stora, came to visit the retirement home where her former professor lived out the last of her days. In 1939, at the age of forty-seven, my mother took advantage of a law which allowed mothers of three children to retire; she stopped teaching with relief. Perhaps her job as a teacher did not satisfy her. She told me one day that it was rewarding to have children because it was a creation. I think that teaching in a lycée did not allow her to give the measure of her capacities.

My mother was very attached to her husband, and did not quarrel with him, at least in my presence. However that did not prevent my mother from complaining to me about him (something he never did to her). In her youth, my mother played the piano. We had a piano at home, but I never heard her play it. She told me she had stopped because my father was not interested in music and would have never come to listen. It was true, indeed, he confessed to me that he was bored at concerts. However, my mother was also gifted in painting. She had painted a few pictures before her marriage, under the direction of her friend Léonie Estrade. She had also given up painting, though my father liked, as his wife did, to see beautiful pictures. I remember, with pleasure, art museums which the family has visited.

My mother had an exceptional understanding of human problems and, as I said before, she lived without mask, never hiding her thoughts. These two facts made her very close to me, but not always easy to live with. I remember some moments which left painful memories. When my brother was five years old he almost died from badly diagnosed appendicitis. He was in hospital after his operation when my mother said to my grandmother in my presence, "If he dies I will commit suicide." My grandmother said, "You have two girls." My mother answered, "They will manage." I repeated these words a little later to Mme Molinier, who came to take news. I am still grateful to her for her answer, "If it had been you, she would have said the same thing."

This answer comforted me, though I knew it was false. Other words which I did not forget are "I don't like to be licked" one day when, arriving for dinner in the dining room where my mother was already seated, I bent towards her to kiss her. However, I loved my mother, she has always been very close to me and has taught me a lot. We really talked together. She made to me, I think, more confidences than to her two other children. Myself, I did not hide much from her. She told me one day, speaking of her three children, "You are the one who is the most like me." This does not mean she preferred me; the truth was contrary. I remember a family outing on a road near Dammartin where, as my mother discussed with my father some problem caused by my sister, I was walking behind them. My mother turned her head towards me and said, "We are more attached to her than to you because her bad temper causes us difficulties." I did not hear the answer of my father, but I found my mother's remark rather bitter. However my mother supported me when I needed it, at least when she was not overwhelmed by grief.

The tragic death of my father dealt my mother a terrible blow; she was angry with the whole world at her misfortune, considering herself a victim misunderstood by her best friends and even by her brother, though he tried his best to help her. Only her son, François, escaped her bitter rancor against everybody. She even reproached me one day for not having participated actively in the Resistance, as if it could have prevented the calamity which struck us. Indeed, my father had warned us, Jeanne and I, against useless acts which he said, to better deter us, "would put our parents in danger". I felt myself heartbroken by the disappearance of my father. I remember the intensity of my wish to find him still in his office one day when I entered at night. The hostility of my mother made me plunge deeper into distress without rebellion, because it is not in my nature and I had not learned yet its necessity.

When our father died, my brother François was only fifteen and I was twenty, but I did not yet have a salary. Moreover, we had to find housing because the Vice-Director of the ENS who replaced my father, the chemist Dupont, wished that we freed the apartment we occupied with my grandmother without delay. My mother wished therefore to resume service, which the law didn't anticipate. The new director of the ENS, freshly arrived from London, where he had valiantly fought for France, seated behind a desk, did nothing to help the family of the man who had defended it, risked his life and lost it. Fortunately, someone in the Ministry, perhaps alerted by my uncle, behaved differently. He found or even had created a position as headmistress for my mother, with housing, for boarding students of the higher classes of the lycée Fénelon that settled in the lycée Montaigne, opposite the Luxembourg garden. Learning of the number of members of the family who had to live there, this compassionate man enlarged the accommodation by transforming part of the attic into living quarters. We were thus able to enjoy a beautiful view of the Luxembourg garden, and an apartment with as many rooms and greater convenience than the one we had at rue d'Ulm. My mother acquitted herself conscientiously of her task of directing the boarding school, and even came to like it, because she had a gift for human relations. Before retiring more than ten years later, she bought an apartment, chosen by her son, in a building under construction on boulevard Pasteur. She went to live there with him when she had to leave the lycée Montaigne and stayed there until the deterioration of her faculties from old age no longer allowed my brother to take care of her.

I am deeply grateful to my mother for teaching me the wish to understand something of this world and for her unwavering support of my desire to learn and to research. She has done her best to help me in the career I have chosen. I think that, being a woman, she has been prouder of my successes than those of my brother, though his were at least as great. I remember the pleasure with which she told me of her conversation with the wife of a deceased professor, whose election in the Sorbonne had awarded me the chair. This spouse who, as it should be, had conscientiously admired her husband, had difficulties accepting his replacement by a woman. Perhaps if my mother had lived her youth in our times of parity, she would have led a different life and been allowed to reveal the best of her abilities. However I remember her quote from a famous woman I don't remember: "La gloire est le deuil éclatant du bonheur" (Fame is the dazzling mourning of happiness.) My mother said to me, a few years before her death: "I had a happy life." I can only be glad of this statement.

4

CHILDHOOD and ADOLESCENCE

Early Childhood 1923–1932

I was born on 29 December 1923 in Lille, the second child, and second girl of Georges Bruhat and his wife Berthe Hubert. It was an involuntary conception ten months after the birth of my sister Jeanne and a disappointment for my parents who, particularly my mother, had wished for a boy. My coming to this world happened at about 3pm, fitting of someone who does not like to get up early. Childbirth was, it seems, particularly painful (the epidural did not exist at that time) because, my mother told me, I was born nose-up in accordance with my scatterbrained character. However, my mother also told me that I was an easy-going baby, suckling willingly and quick to smile, making her feel the joys of motherhood better than my elder of a more difficult character. It was natural that my parents, who already had two girls, wished for the birth of a boy. Following advice found from I don't know where, they practiced, for six months, the Ogino method; used later for contraception, without guaranteed success. It succeeded for my parents; for six months no birth announced itself. My parents tossed the method out and ten months later, my mother gave birth to a boy named François. My mother told me in confidence that the birth of François, five years after mine, had been for her a new revelation, especially when he pissed in her face! My mother loved her three children, but her son enjoyed a privileged status from the beginning to the end. Until I was five years old, I had been favored by my mother and grandmother because of my younger age. I lost that advantage at the birth of François. I foresaw it quite well because when I was told he was coming, I said: "So I will not be the smallest anymore?" However I have almost never been angry with François because of the preference shown by my mother and grandmother. Admittedly, he was overall a nice little boy, and the feminist movement had not existed yet. I played a lot with François until I was thirteen. I then thought it was not dignified to play like a child at that age. I told my brother that I would not play with him anymore. He answered: "I don't care." I was annoyed and pained, I had thought he had enjoyed himself as much as I had in my invented adventures for his toy soldiers, animals or farmers.

I was only two years old when we left Lille, when my father was elected to a Chair of Physics at the Sorbonne. Such a promotion wás very appreciated at that time, when the University of Paris had a privileged status (this distinction between Paris and province disappeared the year I left Marseille for Paris). My mother left the lycée in Lille where she taught literature in "Première Supérieure" (corresponding to a first year of university) for the lycée Fénelon in Paris where she taught philosophy.

During a few years after our arrival in Paris, we lived in rue d'Ouessant, a small street between avenue de la Motte Picquet and avenue de Suffren, near to what was called, I don't know why, "the Swiss Village". Our apartment was on the fifth floor and had a pleasant balcony, six rooms and a long corridor. The first room, near the entrance, was the domain of my grandmother.

In front of it, there was a small room that was given to my sister Jeanne a few years later. On one side there was the living room and dining room, on the other side the kitchen and bathroom. At the end was a storage room and two bedrooms: one for our parents, and another one which I shared for a few years with my sister. She had a delicate sleep and did not want a night-light. I remember night terrors where I thought I saw the devil (who told me about it?) and Jeanne contemptuously telling our parents who had been attracted by my screams, "She believes the nightstand is going to eat her." Jeanne was very happy when she had a bedroom for only herself, after the birth of François, though it was small. I had no problem sharing my bedroom with François.

My grandmother, Louise Hubert born Lainé, came to live with us after the death, in 1925 I think, of her husband my grandfather Eugène Hubert. She was then sixty-three years old. My uncle, her son René, had advised this solution, saying that his mother would help mine to bring up her two young children. It was probably the case in our infancy, though her daughter did not show satisfaction with her mother's care of us because it did not really discharge her of responsibilities towards us. My grandmother, whom we called Mamé, was a discreet and reserved woman, very different from her daughter. The latter had felt closer to her father and her own grandmother, whom I have not known. However, Berthe took care of her mother up to her death at ninety-two years old. My grandmother was not without her strongly-held personal views, full of common sense and without sectarianism, but she did not express them often. She hardly left her room except for meals with the family. However, when my sister and I were young children, she used to fetch us at school and bring us home. She took us on Thursdays, school holidays, to play in the nearby garden "Champ de Mars," where stands the famous Eiffel tower. I remember these afternoons in the park where I ran happily behind my hoop. I would have liked a scooter but my mother considered it a dangerous sport and refused to let me run that risk. I don't remember playing in the park with Jeanne, who was less interested than me in physical activities. Our grandmother sometimes bought us a ride on wooden horses where, holding a stick, one tries to slip rings on it. I was not very successful, perhaps because of my lack of binocular vision, but it was said to be due to a clumsiness coherent with my absent-mindedness. Jeanne succeeded better.

At home I liked to play with Jeanne. I remember the excitement I felt participating in the adventures that I and especially Jeanne, imagined for gold miners represented by pins of games, I don't remember which, in mines made from these cubes with which young children reconstruct images. We also played a lot at mimicking school, the students being pictures cut in catalogs.

We had a maid (in french "bonne", short for "bonne à tout faire" meaning "good to all work") like all families of social situation similar to ours. First it was Olva, the young daughter of proletarians from the north of France. She was courageous and cheerful in spite of her handicap: she had a hip dislocation acquired, she said, by a fall when she was a baby. She limped heavily, but that did not diminish her liveliness, nor did it prevent her from finding a husband. He was a taxi driver, a Russian who got asylum in France after the Soviet revolution. His social origin and culture were well above those of Olva, but he had made a good choice in his wife and they were, I think, happy together. They had three children, and brought them up with love and wisdom. They effectively helped their children to study and integrate into good French bourgeoisie. Olva started very young with us and was attached to our family, especially to my mother. She continued to visit her long after getting married.

Olva was replaced by a young Polish girl, a better housekeeper and cook than Olva — we were not her first employers — but who was more reserved. I have not kept many memories of her, except that she reported the fights with my sister to my parents, demanding a punishment. That was a spanking for both of us, which my father seldom gave but did so with regret. I have not been traumatized by them. I relieved myself by giving long howls. They were not due to the pain, which was light, but to the feeling that injustice was done to me as the attacker was almost always my sister. I was fond of Jeanne, but she did not like me. I think it was partly due to the preference shown to me in my early youth by my mother and grandmother, but also because we were very different. I was eager to please and not too scrupulous in respecting the letter of the truth. Jeanne despised me for that. She had a strong character, respecting what she held true and following her own way.

First Years at School

Jeanne and I went to school only at about seven. It is our grandmother who taught us to read. She had no problem with my sister, but met difficulties with me. I did not show reluctance to come to the lessons, but I did not remember anything. Aunt Marie, a retired teacher, declared during a visit at our home that she would take charge of that task. I was expressing no ill will, but Aunt Marie gave up when she tried to make me read the sentence "Leontine se lave la figure" (Léontine washes her face) surmounted by the ad hoc illustration. Indeed I read conscientiously "Lé-on- ti-ne se dé-bar-bou-ille" (Léontine cleans her face up). Many years later she would still ask with concern, "And how is school for Yvonne?" In fact, I decided to learn to read when François began to speak. I can still see myself in our room thinking, one morning: "I must learn to read, to read him stories." So I learned to read, and quickly enjoyed it.

Our parents put us to school, first Jeanne at seven in 1929, then me in 1930, a few months before I turned seven, in classes which were then called "neuvième" (ninth). They chose a small private school called "Cours Gernez" located at avenue de Suffren, which was easy to reach on foot from our home. It seems that Jeanne got used to it rather fast, but it was difficult for me. I was unhappy, feeling maladjusted among young girls who had already two years of school and whose parents were practicing Catholics, unlike mine. My parents were of Catholic background, especially my father's family, but they did not practice or even believe in that religion. My mother, a philosopher by taste and profession, was interested in religions. She had had a soft spot for Buddhism. As for my father, I heard him say, "Metaphysics makes one crazy." Hence, he did not ignore the existence of metaphysical problems, but he preferred to focus on problems of physics which, at least sometimes, can be solved.

However, Catholic traditions embellished our family life. Baptisms, First Communions, weddings and funerals were traditional ceremonies that gathered the family. Jeanne, I, and later François, went to Mass on Sundays, as was required by catechism, where we went in view of the First Communion. It was indeed necessary to have a signed card attesting our attendance to Mass. It was, in general, our father who accompanied us to church, while our mother prepared the Sunday meal as it was the maid's day off. I think that she did not miss Mass, but my father enjoyed these moments with us, and they reminded him of his youth. If I was not against useless regrets, I would feel some nostalgia for the disappearance of the practice of Catholicism as a tradition that united families, let alone the loss of a common culture which had lasted many generations.

I have two painful memories of my years in Cours Gernez. The teacher had asked that we grade ourselves, after she gave us the correct spelling of a dictation she had done. I had not noticed one mistake and I had given myself one point too much. The teacher noticed it in re-reading my paper and, pointing to the offending word with an accusing finger, she said in a severe and inquiring tone "sciemment?" (a word not much used in French, in English it translates to "deliberately.") I was afraid because I did not know that word. I did not answer, even after she had repeated strongly several times "sciemment?" I don't remember what was my punishment, but I recall my fright at this accusation I did not understand. Another painful memory is a distribution of medals by the priest who taught us catechism. I was the only one to be excluded. Yet I had been a conscientious student. Perhaps my parents' reputation as miscreants were responsible? At the end of the year, the main parish priest came to see us and wondered kindly that I had no medal as he was sure, he said, I deserved it. This kindness was a balm, of which I am still grateful. Among other facts, it reinforces my esteem for the clergy in general.

After three years, my sister left the Cours Gernez for the "sixième" ("sixth", the first year of middle school) in lycée Victor Duruy, that was in our neighborhood and which enjoyed a beautiful park. Lycée (high school) and collège (middle school) were not distinct then; they were the secondary schools. Some schools even housed the first years of study, beginning with the "onzième" (eleventh) up to "septième" (seventh); they were the primary schools. Teaching was not co-educational at that time; Victor Duruy was a school for girls. For logistical reasons I was sent there the same year as Jeanne, in seventh, now called CM2.

At Victor Duruy, we were half-boarders as our coming back at home for lunch was difficult for our parents. At first, I was poorly adjusted and unhappy in Victor Duruy. I tried to join my sister during the breaks, though that did not please her much. Fortunately, we had a teacher of quality, Mademoiselle Blondin, and towards the end of the year I began to adjust. I even made a friend named Simone. However, an event came to change my academic and personal life completely: I was rejected at the examination for entry in secondary school. I still see myself in front of the list of admitted students, looking for my name and pouring bitter tears at not finding it. I had known how to solve the problem given at the exam and found the result which our teacher had given us in making the correction in the classroom. I sent my father to investigate on what I thought to be a mistake. He went and was told that the solution of the problem did not appear in my paper. Before insisting to see the paper, my father went to see Mlle Blondin. She said that I could have been admitted because I had made great progress during the year, but perhaps it was better if I stayed in seventh for another year because I was very young for collège (middle school), since I would not be ten years old at the beginning of term. My parents followed her advice, and my father gave up asking to see my paper.

My Second "Septième" (Seventh)

My parents decided to spare me from doing a second seventh in the same school. They put me, as well as my sister, in the lycée Fénelon. At that time it housed a primary school in its annex of the rue Suger. My mother was a philosophy professor in Fénelon. She was much esteemed there, especially by the head mistress Madame Elichabe. She had many friends among her colleagues.

My life was completely transformed with this change of school. In that seventh I had a teacher, Mlle Perrin, who knew how to make her teaching interesting for me. I liked her. She taught me spelling and grammar. She gave us interesting themes for compositions and made us write comments on literary texts. She was an understanding woman. Once, she asked us to comment on a text where a child received a gift. Her question was: explain the importance of the present. I misunderstood it as us having to explain how the present, used as a tense in the text, made it more lively. Mlle Perrin spared me shame for my misunderstanding. I got good marks right away in French as well as in calculus, and found myself at the head of my class. It was a happy change for me from the three–legged duck I had been at school until then. I gained some independence of mind and confidence in myself which I had lacked before. I was permitted to go back and forth from our home at rue d'Ouessant, to school. For the journey, which soon became familiar to me, I took either the subway to the station Cambronne, or a tramway leaving from Saint Sulpice. I went back home directly, except at times where I would make a stop at the bakery and buy a caramel with a twenty-five cents holed coin received as a reward for I don't know what.

There were, at that time, school supervisors with less diplomas than professors. They ruled the entrance and departure of the students, watched their behavior during the breaks and made sure they respected the regulations. Among those regulations: wearing a gray blouse with our name embroidered in red cotton, interdiction to wear ankle-socks, and various other more or less serious things. One of these supervisors was called Mlle Paquette. She was a character. Single, about the age of my mother, she lived in a hotel. She was very intelligent, a writer and poet in her spare time, but she did not have the diploma necessary to hold the position of teacher. Her work as a substitute was stimulating for young minds. Mlle Paquette had bonded with my mother in a friendship that extended to me and leaves me a warm memory.

The name of my best friend at school that year was Geneviève Limasset. She was the dear little sister of a former student of my mother, who had remained very attached to her former professor and favored my coming to their apartment. Unfortunately the family's mother died tragically that year from a rabbit bone stuck in her throat. Penicillin had not been discovered yet to stop the infection that followed. Geneviève and her sister left Paris and I did not see them anymore. I also had friendly rapport with other fellow students, among them, Micheline Lagier, who shared the "prize of excellence" at the end of the year with me. Micheline chose a career in music and I have lost contact with her.

Later Childhood, Schooling in Lycée Fénelon

I was admitted in 1933 in sixth, that is, collège, without problem. I then left the "petit lycée" of rue Suger for the "grand lycée" rue de l'éperon for seven years between 1934 and 1943, with only an interruption during the "Phony War" in 1939–1940.

I keep a happy memory of my years at school in that lycée of oldish architecture, which had a rather small schoolyard with almost no greens, but where I never ceased to have nice schoolmates. I also found there two remarkable friends who have accompanied me from my early youth until these last years. One of them, Adeline Daumard, has left her mark as a historian. The other, Angélique Lévi born Spitzer, passionate about literature, has evoked admiration and is known for her translations of important novels from Italian and Chinese. Though very different, they were both very near and dear to me, matching, I think, two contradictory aspects of my personality.

Often, I had remarkable women as professors. The quality of the professors in girls' schools was at that time a consequence, happy as far as I was concerned, of the anti-feminism which prevented women from accessing most other occupations requiring a high intellectual level.

In sixth and fifth I was a good student, awarded, at the end of the year, the so called prize of excellence given to the best student. But the lessons in mathematics did not leave me precise memories, and those of French mostly of boredom stemming from the professor. Besides, I think that the French professor did not like me, because she organized a representation of the Corneille play "Le Cid" and had not chosen me as one of the actors. It had shocked the headmistress, Madame Elichabe, who appreciated my mother and knew me to be a good student. She had organized a special part for me: she made me learn and rehearse in her office a satire by Boileau "Les Embarras de Paris". I recited it at the ceremony, wearing a costume of the time and a wig that we had rented. I would gladly have given up the favor, but I did not dare refuse it, and I remain grateful to Mme Elichabe for the time she spent with me on that occasion.

I think we started Latin in sixth, but I remember nothing of it. I was not gifted at languages, and their teaching at school was tedious. I did not like Latin and the frequent versions imposed to us of a few lines of Titus Livius could not arouse my interest. I liked better the study of Greek which we started in fifth. Our teacher, Mlle Duffaut, was born in the southern city of Auch. She had been a friend of my mother, always cheerful, and had known how to interest her pupils. My beloved uncle René Hubert was a fervent admirer of Greek civilization, the cradle of ours, and he contributed to my love of it. He gave me a beautiful bound book, the Odyssey, that was translated from Homer by Victor Berard. This book is one of the few that have accompanied me in my successive changes of lodgings. It is still on my book shelves. I must confess however that I was never enchanted by the reading of the Odyssey. The repetitions annoy me, and there are too many fights and tragic deaths. I continued in the following years to prefer Greek to Latin, though it was now the same professor who taught them; and the Greek translations, a few lines of Xenophon, were now as boring as the Latin ones.

A professor in fifth whom I appreciated was Mme Coquart, a former schoolmate of my mother in the ENS for girls. She taught what was then named "natural history" in a very intelligent way. She asked us to take note of our observations concerning botany, and promised to award a box to collect plants for the best report and a magnifying glass for the second best. I got the first prize, perhaps by favoritism. I was a bit hurt, because Mme Coquart had said ironically to her colleague, my mother, who repeated it to me, "Yvonne wrote that leaves fall". I had noted the dates at which different kinds of trees lost their leaves. Moreover, instead of the first prize, I would have preferred the magnifying glass. The winner, Pierrette Mattei, an independent and original personality endowed with a generous soul, gave it to me while refusing to take the box to collect plants although it was of a beautiful green.

Mme Coquart, née Nelly Fouillet, had held the "Agrégation of physics-chemistry-natural sciences", but she was also a talented writer. Under the name Claire Sainte Soline, she published about twenty-five novels, appreciated by many readers, among whom my mother. One of the first, "Journée", published in 1934 obtained the "prix femina". I read it many years ago. Recently, I read its summary again on the Internet and rediscovered the obsessing realism that I remembered.

An important event for me in my first year of secondary school was the making of my first real friend, Adeline Daumard. We weren't in the same class in the following years because

I was learning English and she was learning German, as was customary at that time for children of intellectuals, especially scientists. Indeed, German science was strong before the war and the persecution of Jews by the Nazis. I should also have been condemned to German lessons. It was the fate of my brother. I escaped it because the Cours Gernez, where Jeanne followed her seventh, had made her begin English. Naturally Jeanne had continued to study it the following years. My parents had thought it wise that I study the same foreign language as my sister. Besides, my father thought that for a girl it was of no importance. I congratulated myself on that later, though when I arrived in the U.S. in 1951 my academic study of Shakespeare did not enable me either to speak or to understand what I was told. My rudimentary but basic knowledge gave me a sure superiority over my unfortunate fellow citizens, like Leray and Lichnerowicz who had studied, in their youth, the language of Goethe. Adeline and I were no longer in the same classrooms but our friendship has lasted and strengthened over the years. Adeline became for me like a sister. She shared my joys and stood by me constantly in the bad moments of my life until her death in 2003, that is, for seventy years.

My first memories of interest for mathematics come from my fourth where Mlle Paule Martin gave lessons on geometry. She must have been a good teacher since she made me love geometry, without my feeling for her, neither her for me, a particular sympathy, though she had, too, been a schoolmate of my mother. She even showed pleasure one time when I did not get the best mark. I did not cease, however, to appreciate her lessons during the two following years where she continued to be my professor of mathematics. She made me love the study of properties in the plane of circles and triangles, as well as the search for solutions of problems clearly posed. I regret that this fascinating chapter of mathematics has, I believe, disappeared from the program of schools, replaced by the premature study of general structures, which are much less stimulating for young minds.

In spite of my love for elementary geometry, the first professor towards whom I felt a sentimental attraction (I reached then my thirteenth year) was in fourth, our professor of natural sciences, Mme Mouktar. I still remember her blonde and curly hair. It was geology which was on the program. Mme Mouktar had awoken my interest in this fundamental science and I started a collection of stones which I pursued for many years, longer than the herbarium I had started the previous year. Stone conservation suited me better; it requires much less care than that of plants. A stone has something immutable, absolute, which aroused my attachment more than a plant denatured by drying. At the end of the year I took the initiative — surprising for me as I am not very gifted for social action — of a collect among my schoolmates to offer a bunch of flowers to Mme Mouktar. I forgot, quickly, this childish devotion. The school year of 1936–1937 saw the birth in me of a different person, an adolescent.

Life at Rue d'Ulm

We left the rue d'Ouessant for the rue d'Ulm when my father became Vice-Director of the ENS. Jeanne and I then had to again share a bedroom that was admittedly very large, though with only one window. My sister had a long table close to it where she worked, a bed and a wardrobe on each side. My bed and small writing table were farther from the window; I had a chest of drawers near to my bed and a wardrobe in the corridor. In the center of the room was a fairly big round table which I usually encumbered with papers and books, to the great displeasure of Jeanne. My parents had thought that their boy François was now too old to share a bedroom with his sister. The apartment at rue d'Ulm was not, in fact, convenient for accommodating our family, but my father was not used to claiming for himself so we contented ourselves with it. I think it has been remodeled after we left. When our family lived there, Mamé had a room of her own that was connected to ours (Jeanne and me), and separated from my parents' by a small room with a sink; the room was given to my brother to use as a bedroom in the first years. Later our parents lodged François in the entrance: it was a pleasant room, but it opened directly to what was called the staircase of honor: a staircase with a red carpet which led only to the apartments of the Director (on the first floor) and of the Vice-Director (on the second floor). Mamé had to cross our room or those of my parents' and François' to reach the corridor leading to the other rooms of the apartment: the dining room, living room, kitchen, and then a waiting room opening on one side to a corridor leading to toilets and a room which was given as a bedroom to the maid, and on the other side, to a staircase without carpet that was shared with the students. The waiting room also opened to a large room which was the office of my father. Behind my father's office was a small room that my parents gave to me later as an office when I was a student in higher mathematics, as well as the unique bathroom of the apartment. It was not without inconvenience to have to cross the paternal office to access it, though my father spent most of the day working in his laboratory. I understand that the vice-director who succeeded my father has demanded remedy to this situation.

However, we had adapted ourselves without pain to this apartment. The building of the rue d'Ulm has a pleasant location:

in the so-called Latin quarter near Luxembourg. It has two gardens, an interior one with a basin that had a pool of goldfish, earning it the name "court of the Ernests", and the other garden on the side which at that time was reserved for the administration. François and I have often played there in our youth with the grandchildren of the Director, Sébastien Bouglé.

After moving to rue d'Ulm, we went to school, usually on foot — with Jeanne and I going separately to the lycée Fénelon, and François to the lycée Henri IV. Our parents did not control our comings and goings. We had only to be home on time for lunch and dinner. We were not allowed out after dinner. Sunday was sacred, a day reserved for the family. When the weather permitted, we took a train to Rambouillet, Chantilly, Fontainebleau or some other beautiful place; there are many in the neighborhood of Paris. We had no car because in those times, the mechanism of cars often required some intervention by their owner, and my father, an experimental physicist by trade, did not like to do odd jobs. He wanted to enjoy his free day without a trivial problem of mechanics. Both my parents liked nature and walking, and we had long walks in forests. Walking did not prevent various games where imagination held a large place. My mother told me later that she would have liked to picnic sometimes, but my father preferred the comfort of a restaurant and, at least on Sunday, good cuisine. We had favorite restaurants and I keep a particularly pleasant memory of a restaurant in Saint-Germain where, during the hunting season (there were no ecologists at the time), my father regaled us with deer, cooked pears and chestnut puree, accompanied by half a glass of Sauternes. I still remember, with emotion, moments when, a bit tired of walking, I clung to the sleeve of my father's overcoat. It seems to me that I still feel its thick and comfortable fabric.

In poor weather, we went for walks in the park of Sceaux, which made my father, who had liked puns, say, "It always rains seaux (buckets) at Sceaux," Sometimes we went to visit a museum or a painting exhibit. On bad weather Sundays, the family often played rami. After adolescence I was not particularly fond of it, but I did not think of escaping the family entertainment.

My parents left their children very free, trusting us in our schoolwork and our attendance to the lycée. However, they, my father in particular, were concerned with our results at school. He made me recite, at my request, the history or geography content to be tested in an examination. He was ready to respond to our request for help for a paper on science or Latin. The French writings were the responsibility of my mother. She read and commented on mine if I wished, which was often the case. She could also have helped me for English papers, since her literary Agrégation was in letters, philosophy, English, but I seldom used her in that domain where I made little progress during my school years. To encourage his children to work, my father kept a little notebook where he wrote our good results and the pocket money they brought us. We had also a small weekly stipend, independent of our results. My father liked that we saved, hoping that in doing so we would learn to manage a budget. However, he never refused when a little later I would spend my savings to buy books. Having neither a TV nor computers, we read a lot in our youth.

The Happy Years

In my youth there was at the end of every school year a ceremony called the distribution of prizes where authorities gave books that were sometimes beautiful, to students who had earned them though their good marks. During the distribution of prizes in 1936 at the end of my fourth, I made the acquaintance of the one who was to become — like Adeline, though very different from her — my lifelong friend, Angélique Spitzer. We were both idle in the playground during a part of the ceremony that did not concern us. Angélique, a few months younger than I, was out of fifth. We engaged in conversation. Angélique, whom her friends nicknamed Anjo, has a remarkable personality, with exceptional intelligence, full of ideas and passionate. She talked to me of her burning admiration for the professor she had had in letters that year, Mme Gateau.

During the summer of 1936, a little before the beginning of term, my sister Jeanne caught scarlet fever and could not leave Dammartin where we had been on holiday. My parents decided that my mother would stay with her with my grandmother, happy to extend her stay in Dammartin, to assist her. François and I would go back to Paris with our father to resume our studies, with Aunt Marie coming to help him keep house and look after us. Aunt Marie acquitted that task very conscientiously, too conscientiously for my taste because my mother didn't use to check if I washed myself properly, nor to come every moment to my bedroom to ask me if everything was alright. I feel some remorse at the impatient tone with which I had answered, "Yes Aunt Marie."

My sister's illness had delayed my coming to school for a few days. I was annoyed because the seats in the classroom had already been chosen and I would have had to go to the back row, which would have spoiled all the pleasure I had in attending classes. Fortunately, Pierrette Mattei had reserved for herself and a classmate, Colette Braive, a table and its bench in the first row. These nice girls offered to make a place for me. I used the pretense of bad eyesight, plausible because of my glasses, to make our professors accept that stretch of the rules.

My friendship extended easily from Pierrette to Colette and I keep an excellent memory of our sharing a bench. Colette was new in the lycée, slightly older than us and thirsty for friendship. Her mother had died several years before. She had been brought up by a grandmother until that year when her father, remarried, had taken his daughter back. Colette did not like her stepmother. It was normal, but unjustified. The latter had only one wrong: caring too much for the welfare of her stepdaughter, like Aunt Marie with me. Colette later changed her feelings towards her stepmother whose kindness had charmed me.

The year of third was for me a complete change. Mme Gateau was an exceptional professor, possessed with a communicative enthusiasm for the great authors of French literature. She made me discover the beauty of poetry and interest in the great literature. Her subjects for French essays led me to think about the great issues of life. Helped sometimes by conversations with my mother, my papers received excellent marks which filled me with joy. I remember in particular the topic "choose your motto", for which my mother had suggested "be free." My writing earned me a ten out of ten. I don't think I have really followed this motto all my life, though the staunch Christian friend I made these last years told me that God created us free: that is how the coexistence of evil and a good omnipotent god can be explained in the Christian religion.

To come back to Mme Gateau, she was a warm and open woman. She welcomed us in her home and shared willingly with us her experiences and personal feelings. She became my idol, as she already was for Anjo. This common passion contributed to our growing closeness. Anjo was surrounded by a small circle of friends of the same class who, like her, enjoyed great freedom. As I said before, my parents also left me quite free in the daytime. As in my childhood, Sunday was always spent with the family. I was surprised when I saw that it was not a universal law. After school our small group often gathered in the apartment at rue des Saints Pères where one of us, Monique Regnier, an unreserved friend of Anjo, lived alone with a rather old father kept outside by his work. Her mother had died years before, and an older sister took care of Monique but was seldom there. Seated on the couch or on the floor we talked a bit about everything: books, paintings, politics and we remade the world. Anjo or I sometimes had the privilege to be admitted in Mme Gateau's home, which was very near, in rue de Lille. Mme Gateau was, I think, happy about the infatuations she awoke in her pupils, probably too much, because that satisfaction cooled them. Mine received a cold shower when its object complained to my mother that we looked at her from the school yard when she was in the teacher's room on the first floor. But why did she put herself at the window if it was not for us to look at her? We were disappointed also, when she offered us her photograph, thus destroying the pleasure we had in trying to photograph her without her knowledge.

The following year, in second, I had the good luck to again have an excellent professor of French, Mlle Broussas. She taught very interesting classes and her well chosen subjects for papers gave her pupils knowledge and a taste for solid classical culture. However, Mlle Broussas betrayed little of herself in her teaching. My attachment went to Mlle Huck, our very competent professor of physics. She had done research in the Curie laboratory, but since parity did not exist then, she had been forced to go into teaching. My father was a physicist and interested in his children's studies, but it was Mlle Huck who excited my interest for that science. She taught it following the method advocated, years later, by my friend and Nobel Laureate Georges Charpak, under the title "la main à la pâte" ("hands to the dough", equivalent to "shoulder to the wheel").

The clear lessons of that twenty-eight-year-old girl of open and friendly looks were very stimulating. The finding of a law of physics, like the heat of change of state, as a result of an experiment done with the help of her advice that gave just enough direction, is an unforgettable experience. The direct and generous personality of Lucie Huck held first my attachment then my friendship until the hazards of life drew us apart. When I came back to Paris I wished to connect with her again; unfortunately she fell victim to throat cancer, perhaps as a result of her work with radium, and had died prematurely. I still think of her with gratitude and affection.

The Year in Poitiers

During the year of the Phony War, 1939–1940, our parents, worried about the announced bombing of Paris, sent their three children to Poitiers where my uncle had offered to shelter us. Anjo's grandmother offered to my parents to receive me in the Auvergne town where she had taken refuge with her grandchildren Anjo and Gérard. Of course, my parents, very family–family, declined her offer. A chance for me was that my friend Adeline, her brother Jacques and their mother also took shelter in Poitiers, with the sister of M. Daumard who was a teacher there. So I found myself in the lycée of Poitiers in the same class as Adeline. We could help each other, though neither of us was permitted to wander after school. Indeed my aunt required, as Adeline's parents had always done, that I come back home directly after class. It seemed to me all the more annoying that, save for a good reason, I had to stay in the room that Jeanne and I shared, though it was admittedly large. Jeanne, already a university student, had more freedom, and François went to play with his cousin Mado. I was too old to play. I did not participate much in the life of the Hubert family, except during meals. My uncle was very busy with his obligations as a Rector. However, I liked my aunt, who gave us a comfortable material life, and I keep a good memory of some interesting and pleasant conversations I had with her.

I did not like the lycée of Poitiers as much as my dear lycée Fénelon, but I don't keep a bad memory from my year of studies in Poitiers. I had there a very good professor of mathematics, Mme Nicolas, exiled from Limoges. She made very clear the reputedly difficult spatial geometry and gave us interesting and appropriate problems for our level as homework. At the end of the school year, Mme Nicolas encouraged me to move towards a career as a mathematician. She told me kindly that she would be interested in my successes. Our professor of physics, Mme Biraud, had been recruited because of the lack of qualified teachers from the war. But she knew it; she was a nice woman and a friend of Mlle Huck, for whom she shared my admiration. The lessons of Mme Biraud were sometimes a bit unclear, but it gave me topics for questions in letters to my father, who always sent me detailed answers. I did not always understand them, but it gave welcome exercise to my neurons and it comforted me with his faithful affection.

Our professor of French literature, Mlle Jobert, was a girl who came from a feminine Agrégation with a program that had been aligned to the one of the masculine Agrégation. It was therefore French, Latin and Greek. Mlle Jobert taught us French literature and Latin. I had been too spoiled by the remarkable qualities of my professors of the two previous years to feel enthusiasm for Mlle Jobert's lessons on French literature, but I liked Latin better than before. Tacite was replacing Titus Livus. I found him more interesting and even easier to understand in spite of his reputation among students. Another teacher, Mme Juliard, was continuing our initiation into Greek. She left me a good memory though it is somewhat vague. I am grateful to her for a good mark in translation in Greek at the baccalaureate.

My favorite teacher during my year in Poitiers was our professor of history and geography, Mlle Desmoulins, a warm and enthusiastic young woman. She made the vagaries of the French revolution and the personalities of its actors exciting. She even made the often dull teaching of geography interesting by linking the various properties of French provinces to their different geologic structures.

At the end of that school year, I did not know whether to choose between a literary or scientific "terminale" (last year of high school). In spite of my reconciliation with Latin, I knew that studies emphasising it were not made for me. I discussed this with Mlle Jobert, a teacher devoted to her pupils and our referent teacher. She repeated to me the flattering words of Mme Nicolas, but understood my reluctance to spend my time frequenting mathematical formulas. We agreed that philosophical studies would perhaps be a good solution for me. I left the lycée of Poitiers in June 1940, and never went back to it.

5

YOUTH 1940–1944

Exodus

During the Phony War, my parents had stayed in Paris. I think my mother had enjoyed being retired and without children to care for. However, when came the defeat and the exodus of 1940, my father had to depart for Bordeaux before the arrival of the German soldiers, to ensure the safety of the apparatus of the brand-new physics laboratory which owed him its creation. My mother, with my grandmother, joined us in Poitiers. I think that due to the circumstances, my uncle and my aunt had not wished to be responsible for their nieces and nephew anymore. That is understandable. My mother, my grandmother, my sister, my brother and I took a train for Bordeaux, perhaps the last before long. In that train, we found Mlle Desmoulins who had sought to join her parents in the neighborhood of Angoulême. A few kilometers from there, the train stopped in open country. An employee of the SNCF told us that the stop was final and advised us to join Angoulême through our own means, that is, on foot. That advice seemed impossible to follow given my grandmother's age. Mlle Desmoulins did not have this problem and was about to be on her way. I went with her as she promised that once in Angoulême, she would send me back to the stopped train in a taxi she knew. I was sixteen, I was a good walker and I had in Mlle Desmoulins a trust that I communicated to my mother. She gave her approval to this project. A few hours later, we arrived in Angoulême where we learned that the train had finally left for Bordeaux. Mlle Desmoulins gave me a bed for the night and left me the following morning in a bus which went to Bordeaux. She had to take advantage of an available car to join her family in a neighboring village. I had been waiting for the departure, comfortably seated in the bus. It had been gradually filled by other travelers for Bordeaux. It was full when the driver arrived accompanied by a woman and a child. In spite of my protests, he forced me to get out to leave them room. I found myself on the sidewalk, alone in a town where I knew nobody. But I was young and brave, perhaps oblivious, and I did not panic. A woman alone and no longer young, also stuck, offered me her support. But she seemed more lost than me and after a polite apology I left her, deciding to hitch-hike to reach Bordeaux. A car soon stopped, and a rather young man was alone in it. He invited me to get in, with a smile I did not like. I declined and walked away. A little later a car containing three or four soldiers stopped. I explained my wish to join my father in Bordeaux. It was their destination. The driver, who seemed to me a trustworthy man, accepted to take charge of me. About mid-way he said, "We could stop on the way and sleep in the hay." The other passengers did not answer. But I had understood he was teasing and I said without emotion, "Why not?" In fact this military man took me directly to Bordeaux and even, as I had told him I hoped to find my father at the university, brought me there and got out of the car to inquire about the presence of M. Bruhat. By luck, my father was on the stairs in front of the building, and the army, represented by the soldier whom I had trusted, left me in his hands before resuming its journey.

The remainder of the family was already in Bordeaux. Colleagues of my father had found lodging for us in a house at Caudéran that had been abandoned by another colleague. The most vivid memory I have of that stay is of the rug in the bedroom I shared with my sister. That carpet was covered with fleas which jumped on our legs as soon as we sat on the edge of the bed. We took it out in the small courtyard near the house. Our stay in Bordeaux-Caudéran was short because the armistice signature was rapidly followed by the return of the whole family to Paris. None of us were happy about the armistice signed by the Maréchal Pétain, nor about the Vichy government ascending to power. My father could have left for England with the last boat leaving Bordeaux. However, having passed the age to fight in the army, he thought his duty was not to abandon to the Germans the modern apparatus of physics he had taken from Paris, but to bring them back, take again his responsibilities at the direction of the ENS and do his best to protect students and material against the German enemy. He did exactly this, which cost him his life in a deportation camp at the age of fifty six. I will speak of these events later, briefly because I cannot evoke their memory without feeling again all the pain they caused me.

Return to Occupied Paris

When coming back from Bordeaux in June 1940, we found Paris intact; the German army had entered it without fighting. We can only be glad of that today. Life resumed its course under German occupation. My parents strove to make it as easy as possible for their children. They did their best so we did not suffer too much, neither physically nor morally, of the presence of the occupant.

They showed confidence in the victory of the Allies and spoke little of the tragedies that surrounded us. However, my parents, especially my mother, had many Jewish friends and worried for them. My father was head of the ENS, first because the Director Bouglé was dying from cancer, later because his replacement, Jérôme Carcopino, appointed by the Vichy government, had left his position to become Minister of Education of the same government. In normal times, my father, a scientist and the Vice-Director, would have succeeded to the literary Bouglé, but he did not please the authorities of the time. Sure, he was opposed to their politics. He did his best to protect the Jewish students of the ENS. The law required their dismissal, and soon several were deported to death-camps. Some of them for whom my father found a refuge in the so-called free zone (the Germans took possession of the south of France only in 1942 and could not control it so strictly) told me, after the war, of their gratitude towards having been saved by him. To protect their family, he absented students, citing illness, who then escaped France across the Pyrénées and enlisted in the English army. My father, however, always dissuaded us — my sister and I; my brother was only a child — from taking useless actions of resistance which, he said to better dissuade us, would put the whole family in danger.

Soon after our return to Paris, I passed the first part of the baccalaureate, which I had prepared for in Poitiers. It was reduced to the written part, organized in haste. If I remember correctly, my best mark was — oh surprise! — in translation from Greek. I was admitted, no distinction attributed.

At my request, my parents enrolled me in the lycée Fénelon, in scientific "terminale" (last year of secondary school), called Math-elem, with additional lessons of philosophy, so I could obtain, with the scientific baccalaureate, a literary baccalaureate, thus delaying the necessity to choose between these two channels.

In Math-elem, I had a very disappointing physics teacher, who failed to disgust me of that science, and a quite remarkable professor of mathematics, Mme Chabauty. She was a widow from the First World War who had lived alone with her son. He had entered the ENS and defended a thesis in mathematics. He was a professor of university, I believe in Grenoble. Mme Chabauty was proud of her son and often mentioned him to us as an example. I must say, to her credit, that she got along very well with the young woman he had married. Mme Chabauty loved mathematics with a communicative love. She got interested in me. She gave me additional problems to solve, and rebuked me sometimes for not working more. I think she would have liked me to earn a prize in the General Competition which was open to all Math-elem students. That success would have made her teaching qualities public. Unfortunately it did not happen. I thought I had completely solved the problem given to us at that competition, but I draw poorly. Hence, my figures were unclear, especially because I had forgotten to take an eraser. Anyway, copies of other competitors were judged better than mine. On the other hand, I got the second prize at the physics General Competition, the first woman to receive this distinction. My father was very happy and the announcement he made to me of this news remains one of my greatest joys. Mme Chabauty did not manifest any rancor, but she told me, "Yvonne, you had a prize in physics, but you are made to do research in mathematics." This statement reassures me when I doubt my having followed the path best for me.

I had found, in Paris, my two friends. Adeline, now in my class and enrolled like me in the philosophy option, shared my table in the first row. Anjo, also back to Fénelon, prepared her first baccalaureate. Her remarkable intelligence earned her the second prize in French composition at the General Competition.

We were both happy about that. Strange as it seems, Anjo had a poor rating in French composition at her first baccalaureate. She was admitted to it anyway, even though her brilliant mind was shut to mathematics despite my unsuccessful attempts to make her accept its logic. Obeying the laws of the Vichy government, she wore the yellow star required for Jews. This did not prevent us from walking together in the streets of the Latin quarter where, I must say, the occupants were seldom seen. Her friend Monique had even sewn a yellow star on her coat, bravado which I thought unnecessary and that my parents would not have tolerated.

Our small group of friends was again meeting in Monique's apartment for long discussions on all subjects, confident in the future. I particularly remember the small party (nothing in common with today's parties) we had enjoyed to celebrate my eighteenth birthday and the gifts, simple books, that these friends had given me on that 29th December, 1941.

A thunderbolt in our small universe was the Vel' d'Hiv Roundup where Anjo's father was arrested in his apartment, at place de la République, then deported to Auschwitz. One did not know, then, the tragic destiny awaiting the deportees. Fortunately for him and his family, Spitzer was a doctor and he spoke German fluently. These two facts allowed him to come back alive from deportation.

Choice of Career

Entering my last year of high school, I had not chosen what to make of my life. The lessons of philosophy I had followed did not encourage me to continue in that direction. It was a good exam preparation, but rather boring. In my memory, it was always "the pros, the cons, conclusion: the middle ground".

Moreover, I had noticed how painful for me it was to write a philosophy paper, never succeeding to express completely exactly what I thought. On the contrary, looking for and writing up the solution to a mathematical problem was a real pleasure, because it involves coming to the exact formulation of a definitive truth resulting from precise hypotheses. I had been admitted to the literary baccalaureate with the distinction “good”, not “very good”. A regurgitation of the lesson given by our philosophy professor on “the idea of object” had earned me 16/20 at the writing part. But my very scholastic choice for the oral part, “the emotions according to William James” rather than “Plato’s Allegory of the Cave” with the examiner encouraging me to take the second, had sent down my mark to 14/20.

In Math-elem, like in the previous year, I had excellent professors of history. The first was the friend of my mother who liked me, Mme Molinier. She was forbidden to teach after one trimester: her Jewish origin had been revealed to the Vichy government by her maiden name, Kont. After the war the director, Mrs. Elichabe, was reproached for the ostracism of Jewish professors. But she could do nothing about it except resign, which would have made the situation in the lycée worse. The successor of Mme Molinier was a young woman whose name I forgot, but she was also an excellent professor. At the baccalaureate I had the good luck to be given a subject which she had treated the previous week. My repeating of the excellent lesson I had heard earned me a nine over ten. The examiner, a man with humor, gave me that good mark in spite of my answer to his geography question: “Where do the Canada apples come from?” It had been: “Probably not from Canada since you ask me.” I was very interested in history, but I was hesitant to make it my profession. Indeed, for a long time I had wished to do research like my father rather than teach like my mother who, though a very good professor loved by her pupils, had not found happiness there. I hesitated a time between mathematical physics and history, but I had the intuition that historical research required qualities which I did not have. My friend, Adeline Daumard, chose this option. She became an internationally known historian and professor at the Sorbonne. Her research in the notarial archives introduced an original method for societal studies. She found new and important results on the history of the XIXth century French bourgeoisie, in particular Parisian. One can be convinced of that, by reading online the reproduction of an article written in her memory by her former student J. P. Blay for a volume on urban history. However, speaking with Adeline of her work, I congratulated myself for not choosing that option to which she devoted so many hours of work seeming to me to have many unrewarding aspects. I don't think I would have been a good historian, though I am still interested in the reconstitution of historical facts by others.

Sciences tempted me more than letters for a profession. However the universe of science seemed to me far from life. In fact, my true wish in finishing Math-elem was to become a doctor. I had read, with great interest, the book on children's medicine that my parents owned. Uniting scientific knowledge and its applications to human beings seemed to me more tempting than a life among equations. I communicated my wish to my parents. Both discouraged me, not really because the Vichy government had suggested that the profession of doctor should be forbidden to women: they believed in the Allied victory. A motive for my mother was the bad memory she had kept of the woman doctor who had nearly caused the death of my brother when he had appendicitis. As for my father, he thought that a medical profession, very demanding, was not compatible with the life of a mother, a normal destiny for a woman. Moreover, he said, he could help me if I chose a career in sciences, but not if I decided on medicine, an environment where, he thought, it was only possible to have an interesting situation if you had personal support. He suggested that since I had an interest for what was then called natural sciences, I direct myself to the study of one of these disciplines. He gave the example of a geologist woman professor at university, a rare occurrence at the time. She had just come back from an expedition — to Madagascar I believe — with a trunk full of stones and a husband, a significant acquisition. Indeed an important wish of my parents was that their daughters found good husbands. It is true that I had shown interest for botany, then for geology and finally for biology. When my parents had me operated on for my strabismus — without much success unfortunately — they had proposed to buy me a jewel to console me. I had said that I preferred a microscope. They were surprised and gave me a jewel anyway, a pretty brooch, and also a small microscope which I kept until I gave it to my daughter Geneviève. My father took me to the laboratory of the ENS where a famous biologist, Robert Lévy, worked. He showed me kindly how to prepare slides to make observations with my microscope. I made a few, but I am neither very skillful nor patient, therefore not very gifted at experimental work. Besides, I was not tempted by studies in the natural sciences of the time, which seemed to require a lot of memorization. I was more attracted by theoretical physics or astronomy. My father was not enthusiastic about these subjects. The works of the head of theoretical physics in France at the time, Louis de Broglie, seemed to him rather far from reality. As for astronomy, he told me, "It is astronomic calculations", a familiar expression deserved at a time when computers did not yet exist. It is said that Gauss, a brilliant mathematician, wasted years of his life making such computations. Finally it was agreed that I would do a year of "Mathématiques Supérieures", called "Hypo-taupe" in the lycée Fénelon. There I would prepare for the competitive entrance examination for Sèvres (École Normale Supérieure for Girls), the best road to science careers. If I really did not like it, I could turn to medical studies.

Jeanne's Marriage, Hypo-taupe and Taupe

Our parents decided that my sister Jeanne, apparently less robust than me, would not prepare for the competition for Sèvres but would continue her studies in classical letters (French, Latin and Greek) at the University of Paris. She agreed willingly and was admitted to the BA degree easily. To prepare for the "Agrégation", my father obtained permission for his daughter to follow the lessons given at rue d'Ulm to ENS students. They did not like this favoritism and decided not to speak to her, except for one of them, Pierre Reboul, who thought that conduct unjustified and treated her amiably, so much so that they married in the spring of 1943. Our parents were reluctant because during the war, Pierre, probably because of the ordeal, had developed type 1 diabetes, a heavy handicap. Jeanne had a strong character and obtained what she wanted anyway. The wedding took place in the Saint Jacques church. They had four children, two daughters and two sons. Jeanne educated them firmly. They all became important officials in our society. Jeanne has been a professor in the lycée of Roubaix until her retirement. Pierre died relatively young at seventy years old, but he had a full life. He was a good husband and a good father, a professor and later the Dean of the University of Lille. I liked my brother-in-law. I had friendlier relations with him than with my sister. He told me one day that if Jeanne did not have more affection for me, the fault was in the possessive character of our mother which did not promote good relations between her children, to say the least. I think he was right. Later in her life Jeanne became nicer to me, though we were never very close.

As for me, in 1941, I started to prepare for the competition for Sévres. During my year of Hypo-taupe, 1941–1942, I had interesting professors. The philosophy lectures were given by a man, M. Mouy, who in lieu of philosophy read to us excerpts from the novel "La Chronique des Pasquiers" by Georges Duhamel. I went to speak to him at the end of a lesson to ask for him to teach us philosophy. This demand seemed to surprise him and made him laugh. However, afterwards, he gave us some interesting lessons on Spinoza. In mathematics, Mlle Leconte, and in physics, Mlle Pompei, without having the charisma of Mme Chabauty or Mlle Huck, were giving us teaching stimulating for the mind. Mlle Pompei, like Mlle Huck, had worked in the Curie laboratory (though longer than her) before turning, by force, to teaching. Her lessons were dense but very clear. Both gave me high marks which convinced my parents and myself that those studies were the road I should follow. I cannot blame them now. For a long time I had some regret at not having chosen medical studies, but as has been rightly pointed out by my husband Gustave, I would perhaps have been just an ordinary doctor, and not a member of the Academies of Sciences. I particularly appreciated, since my retirement as a professor, the mathematical researches on problems linked to physics. They allowed me to never get bored.

Life went on for our small group of friends, for some months. Anjo had passed her philosophy baccalaureate in Paris and did a brief trial in medical studies, to follow her parents' example. But the first year, involving essentially mathematics and physics, was not to her taste. Besides, her mother was a very sensible and able woman. She assured the survival of her children and herself by leaving Paris for the so-called free zone that was not yet occupied by the Germans. They took refuge in Grenoble and lived there underground until the liberation. I had news of Anjo only episodically but we never lost contact completely, and we found ourselves just as close every time.

The following school year, 1942–1943, I was in the so-called "Mathématiques Spéciales" or "Taupe". It has not left me such good memories as previous ones did. One prepared there the admission to the "École Normale Supérieure for Girls", then called Sèvres because of its geographic location. The best friends I had the previous year in Hypo-taupe, Marianne Equios and Claudine Roklawski, had left the lycée — the first, to study pharmacy, the second, for a first year of university (it was called "Math-géné"). On the other hand, those of my fellow students who went to Taupe were joined by an important number of girls who repeated the year to pose another trial for the competition. In the math class, taught by a man, reigned an oppressive atmosphere of preparation to the competition which unpleasantly replaced, for me, the pleasure of learning I had known until then. I made my father promise that if I was not admitted in Sèvres, I would not do the preparation again, but would continue my studies at the university like a number of my previous fellow students. My father promised it. However our preparation, too academic for my taste, had been efficient and I was admitted. I had done very poorly at the second written test on mathematics: to overcome fatigue during that last exam, I had taken, on medical advice, an amphetamine which had made me unable to concentrate. I did better at the oral tests, particularly in physics. I passed the competition as "cacique", that is top of the class.

I wished to do research in theoretical physics, because I had known I was not very gifted for experiments due to a lack of skill and overall, of patience. Mlle Pompei advised me that if I wanted to do such research, I should learn a lot of mathematics. It was good advice.

First Years in the ENS for Girls

In October 1943, I became a "Sévrienne", that is a member of the "École Normale Supérieure" for girls. The name Sévrienne comes from the small city in the neighborhood of Paris called Sèvres, where is located the beautiful building which housed the "Normaliennes" before the war, surrounded by a park.

This school was not, at the time, merged with the ENS of rue d'Ulm, reserved for boys. The next step in higher education, the "Licence" (bachelor degree), was, however, and we prepared it with the boys at the Paris university, called then the Sorbonne. Indeed most of the superior teachings, sciences and letters still took place in that venerable building. Only a few additional courses were given separately to male and female students of an ENS, in their respective residences.

The building that housed the Sévriennes before the war had been requisitioned by the occupying forces. The Ministry of Education had found, for us, two houses in Paris, left vacant by the departure of their owners. Both were originally American cultural centers. One of them, on boulevard Raspail, lodged scientists. The other one, located not far from the first, on rue de Chevreuse, was assigned to the literary. We all took our meals there, which enabled happy interdisciplinary contacts. The location near the Latin quarter was actually more convenient than the suburbs to allow the Sévriennes to continue their study at university.

The two years following my admission to the ENS were somehow incidental regarding my studies. The Sévriennes and the Normaliens of rue d'Ulm followed the same courses and had the same exams as the other candidates to the Licence of Paris University. Mathematics classes took place near the Sorbonne, at the Institut Henri Poincaré on rue Pierre Curie (the name Marie was added only later). The class on analysis was given by Valiron, a competent mathematician but a boring teacher who had no personal contact with his students. He had written a very precise and complete book that he followed exactly. I remember hearing him say at the beginning of a demonstration: "Let us set d=epsilon/192" to find, after a long sequence of calculations, a theorem where epsilon appeared. The geometry course was given by Garnier, a much more pleasant man than Valiron, but his classes, given to passive students, did not offer incentives to personal work. There were, at that time, no assistants to propose exercises, or submit and correct problems before the final examination. Garnier did not follow a book, but his course, identical to the one in the preceding year, was available in duplicated copy. My only motivation to attend classes was to meet classmates. I found it more fruitful to study the relevant subjects in books. We, Sévriennes, had a lady who watched our good conduct, Mlle Descoustal. Finding me in my bedroom one morning she asked me reproachfully, "Yvonne, why are you not at the university?" My answer: "I stayed to work", left her speechless.

The course entitled "General Physics" had an "Optics" part taught by Jean Cabannes. His lectures were clear and informative without being exciting. The other part, "Thermodynamics", is a delicate subject on which our professor, Edmond Brun, failed to capture my attention. I was chatting with my neighbor when Brun, displeased, called me and asked me my name. I fled without giving it to not dishonor my father.

After the work done to be admitted in the ENS, that required by the Licence was light. We all passed the examinations in mathematics and general physics, without excessive effort.

The supplementary classes we received in Sèvres had little success with students. Garnier gave a complementary course I almost never attended. I remain grateful to Garnier for his kindness towards me: I was with my father rue Saint Jacques when we met him. He introduced me to his wife, who accompanied him, as, "Yvonne Bruhat who is my brilliant student in Sèvres", when I had cut almost all of his lectures! After the death of my father, Garnier was always kind to me. He even invited my husband Léonce and I to dinner in his home. Other lectures in Sèvres were given by the famous mathematician, Elie Cartan, who was seventy-five years old. His course, on some peculiar properties of some sets of spheres, was difficult to follow. By the third lesson, no student was present. Cartan went to see our scientific director, Mlle de Schuttenbach, to complain. I was summoned in the directorial office and asked, as a "cacique", to go to the Master's home and present my apologies to him. I obeyed and rang at the door of Cartan. I had prepared an excuse of the kind "we had to participate to …", I don't remember what I had invented. Cartan interrupted me on my first words, saying, "Do not apologize. I have understood that I have to give my resignation. I am too old, I think in the subway of what I will say, and when I am in front of the blackboard I don't remember." I recall this lesson in modesty now when I receive an invitation to give some lecture at a congress.

We worked on experiments in physics, in the new laboratory built at rue Lhomond just before the war essentially under the direction of my father, with the financial support of Jean Zay, then Minister of National Education. The program of studies having become the same for girls and boys, Georges Bruhat had authorized the Sévriennes to come to work in this newly equipped laboratory. I don't know if he had foreseen that his daughter would have gone there to do experimental work, rather clumsily in fact. My father did not teach for the Licence when I prepared it, but he came to the physics laboratory when we were working there. I remember one day when I had done, with great effort, the setting of some apparatus; my father, passing nearby and wishing to show me something, touched and destroyed my setting. I let out a horrified moan; in one quick gesture he redid the setting that had taken me nearly an hour. I understood what I had already intuited, that I was not made for experimental physics. Some physicists have told me since that to experiment now, one has only to push buttons. But I doubt it and I don't regret being satisfied in believing the results found by experimenters more gifted than me.

Private Life

I was happy at home between my parents and my brother. I would gladly have continued living with them. My father dissuaded me, saying that Normalienne I was, as a Normalienne I should live. Of course I would come back for the week-ends and could go see them as often as I would like. I followed his advice, and was happy for it. Life with girls my age and similar interests proved as happy as possible, more bearable than it would have been without their company when problems and disaster came for me. We were sixteen in our class, there reigned an atmosphere of great solidarity, perhaps especially thanks to the fact that some of us had been brought up with a deep respect of the true values of Catholicism. However, this solidarity extended to the whole class, and still existed fifty years later when the class met for its fiftieth, then, with the number unfortunately reduced by deaths during its sixtieth anniversary. I was a believing and practicing, Catholic when I entered Sèvres. Ever since I was very young, my mind had been tormented by metaphysical problems.

My admiration for Mlle Huck, herself a recent convert, my encounter with a priest, the Father Brillet, who shone a contact with "something else", had led me to believe, pray and attend morning Mass. I would have difficulties now understanding how it is possible to believe in a religion without losing the power of logical reasoning, if I did not remember that period in my life. It helps me to admit that we create our own reality. Christianity was part of it for me for some time. However, I have been somewhat disappointed by conventional Catholicism as practised by most of my comrades; though only the result of their education, they did not seek further than the letter. It was not the case for one of them, Elizabeth Germain who, once Agrégée of mathematics, became a nun. She refused to teach mathematics in a Catholic institution and resumed her studies, but in theology, which she then taught.

In fact, I became especially good friends with two miscreants, Marguerite Barthès, called Maïtou, and Nicole Fontanel. Maïtou had a warm heart and varied interests. She had had a very happy childhood, the only child of two parents who loved each other. She loved them both, especially her father, who was a teacher like my grandfather. She liked to talk about them. She already had true friends, the way I did with Adeline and Anjo. One of them, the daughter of a Communist Member of Parliament, came to visit us in her father's function car with a driver in white gloves. Nicole was more reserved; I believe we were her first intimate friends. Perhaps because of that, I think Maïtou was more attached to her. Nicole was tall and slender, pretty and perfectly elegant, physically and morally. During our first year in Sèvres, there had been a lack of rooms which had forced the administration to put two students in the same bedroom. So we had to gather in pairs. The class decided that Nicole and I would be well-matched, and we had readily agreed. Nicole was always perfectly kind to me. She never seemed to notice the untidiness and lack of care reigning in my closet that had caused my sister so much suffering, and which had made a striking contrast with the perfect order of hers. Nicole, Maïtou and I were a trio of different personalities, but we understood and brought one another unwavering support, in our work as well as in the hazards of our sentimental lives.

Daily life was pleasant. The housekeeper, a woman full of resource, succeeded in finding enough to feed her flock in spite of the restrictions of the time. Sometimes it was food reserved in other times for animals, but I don't remember having suffered from hunger like many of our less fortunate fellow citizens. Meals were served in rue de Chevreuse for all students, literary or scientific. Many members of the class, very united, took their meals at the same table. It could not accommodate us all. In fact, Nicole — of an independent nature, and Maïtou and I — wishing to broaden contacts, lunched or dined most often together at one or the other table where seats were still vacant. I thus made two friends in the literary class contemporary to mine: my "sister", that is the nice "cacique" Claire Gouton, and the philosopher Madeleine Biardeau, who had attracted me with her interest for metaphysical and religious problems. I, unfortunately, lost contact with them after the overturning of my life. I regret it, especially for Madeleine Biardeau, who has become an outstanding Indianist, learning Sanskrit and publishing, among other works, a translation of the voluminous book of Hinduism, the Mahabharata.

However, until her death a few years ago, I had the chance to pursue the friendship I had made in Sèvres with a schoolmate who was a year ahead of me, Madeleine Lebon. Madeleine had passed the scientific competitive exam, but once at the ENS she had decided to follow studies in philosophy, which was allowed by the rules. She hence did what I had considered and that brought us closer. Madeleine Lebon was a full believer, brought up in a practicing Catholic family where she was happy. She was also very close to her brother and sister. "I have never doubted", she told me one day. My mother pointed out to me that that assertion is not very compatible with the profession of philosopher. Madeleine was, however, a very intelligent woman who had passed, with "very good", both the scientific and literary baccalaureate, a feat I had not managed. As the Agrégation of philosophy did not offer many positions, she had not been admitted to it. She had considered a re-conversion to mathematics — my entire class had been successful at the competitive exam — when marriage and children convinced her to remain a housewife. Fortunately for me, Madeleine married a mathematician, Henri Cabannes, the son of a Normalien of the same class as my father as well as his friend and colleague at the Sorbonne. Henri, a year older than me, had passed the ENS entry exams a year before I entered Sèvres, but he had fled France through the Pyrenees at the risk of his life to enlist in the air force and fight Germany. The Cabannes, and Henri at the time, were practicing Catholics. Madeleine and Henri met at the Catholic youth gatherings, liked each other and wed. They remained my friends our whole lives. I will speak more of them later. Henri told me, towards the end of his life, that he had lost his Catholic faith because his wife was too much of a believer!

Breakfast, like other meals, was served at rue de Chevreuse, but a small room with a gas stove allowed us to make hot drinks in our building. Maïtou, an early riser and a devoted soul, often went to rue de Chevreuse to get the solid food; we made tea with the stove at our disposal. After that, Maïtou, Nicole and I had breakfast in one or the other of our rooms, speaking about work or other subjects.

Sévriennes and rue d'Ulm Normaliens of the same year met often. Some of us, coming from the provinces, had been in mixed preparatory classes where they had already made friends. The boys housed in dormitories at rue d'Ulm, with four to an office called "turne" where they spent the day and worked. This facilitated the fostering of friendships, and sometimes led to weddings. This is how I met Léonce Fourès whom I married in 1947. He shared his turne and was friends with Emile Pallarès, a very likable Normalien who had been with Maïtou in the Taupe of Montpellier. The two other "coturnes" (roommates), Jean Rogues and Albert Monestier, were also pleasant relations. Emile and Albert became Taupe professors: Jean wanted to have an ecclesiastical career, and perhaps become a bishop. He became a parish priest, in the beautiful church of Saint Germain des Prés. Since his retirement he has been the president of an ecumenical group. Léonce has been the only one in his turne to have a career as a mathematician and university professor. Other mathematicians of our class had careers in research. Michel Parreau, cacique of the class, has been a professor, then the Dean, of the faculty of sciences of Lille. André Néron did important works in algebra; René Thom, after introducing in mathematics the very useful notion of cobordism, has become famous with his "catastrophe theory". Both have been members of the Académie des Sciences. I had very friendly relations with Parreau and Thom, who were also open-minded about problems outside of mathematics.

Maïtou had a tender spot for Emile, who returned only a great friendship. The generous Maïtou did her best to bring him close to Nicole, with success since he married her, supplanting another suitor of that tall and beautiful girl. Maïtou herself got married soon after leaving school, to a former student of "école polytechnique", the son of friends of her parents. I am happy to say that this marriage, partly arranged by the families, has been particularly happy. For my part, I did not have with the other sex the romantic success that I wished, probably because of my strabismus. Nicole told me kindly that she received more attentions when she removed her glasses, but removing my glasses would not have helped. I was touched by the kindness of Emile's coturne, Léonce Fourès. He was the single child of professors, his mother of natural sciences, and his father, a violinist and professor of music. At the time, Léonce had been exclusively attached to his mother and had not played any instrument. I heard that later he reversed his preferences: in a speech for Léonce's funeral one of his former students said that he was a violinist and had loved his father. I can believe that his feelings had changed, it is supposedly common, but I have a hard time believing that he started playing the violin after thirty and became a master of that difficult instrument.

Emile and Maïtou had become friends in Montpellier with a classmate who was then known as Georges Charpentier, hence hiding his Jewish origin, which the name Charpak would have betrayed. Charpak's family had fled Paris and taken refuge in the free zone, having been warned early enough of an upcoming raid by one of Georges' classmates whose father was a policeman. Georges was, however, arrested by the gestapo in 1944 and deported to Germany, fortunately not as a Jew, but as a Communist and resistant. Maïtou had given moral support to Georges' mother, convinced that his robust youth and his courage would bring him back safe and sound at the liberation. Fortunately for his family, friends and science, she was right. Georges told us, after his return, that his survival owed a lot to the solidarity that reigned between "comrades", that is, between young Communists. Georges waited a few years before "getting tied down" per the saying, but he found a partner who suited him and had with her three children, two boys and a girl, who are all listed on Wikipedia. I often met again with Georges who had a brilliant career as a researcher. He was elected at the Académie in 1985 and received the Nobel Prize a few years later. His books, including his autobiography "La Vie à Fil Tendu" made him known outside of the scientific field. His foundation "La Main à la Pâte" widened his celebrity status. A dozen streets in various French towns now bear his name.

6

DISASTER 1944–1946

Deportation

The summer of 1944 saw the liberation of the French territory, but for me, it was the year of the disaster: the arrest and deportation of my father. The Director Carcopino, maybe because he foresaw difficulties following the landing of US troops, had chosen that moment to take his holidays. He informed my father that consequently, he must remain in rue d'Ulm where there still were many students unable to leave Paris. My father then wished that my brother, myself, my mother and my grandmother would go to Dammartin where he would join us later. My mother refused, citing the difficulty of the trip for my grandmother. Thus we all remained in rue d'Ulm. The gestapo landed there one evening, looking for a student belonging to the Resistance who had just fled. The gestapo ordered my father to find him and took, as hostages, my mother and Mme Baillou, the wife of the General Secretary, a literary. My father did nothing to find the sought student, but sent us away for the night: my brother and grandmother at an old friend of hers, Julia Lohmond, and me at the ENS for girls. When I came back for news the following morning, I learned that my mother and Mme Baillou had been released, but their husbands taken. It was a few days before the Paris liberation by the General Leclerc. The gestapo put its prisoners in a train to Germany. The US army dragged to end a murderous war, and only the particularly robust young deportees saw France again. My father did not come back, his ashes are in some common pit beyond the Rhine. Baillou, only in his thirties, came back alive from deportation. I saw him a few weeks after his return, a walking skeleton.

Waiting with Hope

We had not known until the liberation of the camps the horror of the treatments inflicted by the Nazis on the deported. I was twenty, I was confident in the Allied victory and the return of my father. My mother also did not imagine the horror of concentration camps. She was worried by the nomination of Pauphilet, back from England, as Director of ENS, thinking that the position was due for her husband. However, she encouraged me to approach priests, whom I knew through my conversion to Catholicism, to try to get news of my father and come to his help. I went to see the chaplain of the Normaliens. He asked me, "Was your father practicing?" Not having thought of having to lie, I said no and the priest said, "Then I will pray for him." This response began to shake my faith. I took advantage, a little later, of a conference held at our ENS by Father Danielou to ask for his help. His response was not that of the previous priest. He gave me the address of a woman who was active in helping deportees. I went to see that lady with his recommendation. She was very understanding, shocked that the gestapo had dared to arrest a scientist of the value of Georges Bruhat. She told me she would do her best to get a package to him and even try to have him released. I brought her the package including our best blanket and we hoped for news which did not come. However, we always hoped for his return. I learned later that he had been transferred from Buchenwald to Sachsenhausen, which had made locating him more difficult.

During the summer of 1944, I prepared, with Léonce, who remained, like me, in Paris, a certificate of "General Mechanics", using the very clear book of Chazy. We both passed the exam with success. I had been questioned at the oral exam by Henri Cartan, who knew nothing of mechanics. He had given me, as an exercise to solve on the blackboard, the study of the motion of a punctual weight on a vertical circle. I found, as it should be, a motion in general oscillating except in two points of the circle, the highest and the lowest, where when the initial velocity is zero the weight remains immobile: they are called equilibrium positions, stable if the weight oscillates in their vicinity for small initial velocities. Intuition, that comes from experience as any intuition, says that the highest point gives an unstable equilibrium and the lowest a stable one. I had gotten confused in my notations and found as stable the highest position. I was ready to acknowledge my mistake when Cartan said forcefully, "Very good, Mademoiselle, very good." I passed with distinction. A very competent mechanics physicist, Joseph Pérès, asked Léonce a delicate question on tides, which got him only the second class honors. This injustice caused Léonce some bitterness.

In October 1944, we were both licence holders. We lacked only a diploma of so-called "superior studies" to compete for the Agrégation the following year. Traditionally, Sévriennes prepared for the certificate "Celestial Mechanics" taught by Chazy. I went to the first lesson, but the procession of equations written on the blackboard had dissuaded me from persevering. It is amusing to think that seventeen years later, I would then be elected at the Sorbonne in the Chair that had been occupied by Chazy.

Still wishing to become a researcher in theoretical physics, I chose to prepare for the certificates "Probabilities" taught by Georges Darmois and "Superior Analysis" taught by Montel. The content of the chosen certificates was light. The definitions given on probabilities were intuitive and the theorems easy to prove. The superior analysis was an elementary study of almost periodical functions. To pass these exams required little work. Darmois had been a schoolmate and friend of my father at the ENS. He told me one day, "Your father was hardworking, he became a physicist. I was lazy and that is why I did mathematics." In reality he did fundamental work on General Relativity and later, I think, in statistics, but he liked to change topics after groundbreaking work. He was always kind to me.

In the fall of 1944, the Sévriennes had been satisfied of the departure of the Director, Mme Hatinguais, a literary who considered Sèvres a boarding school for young middle-class girls who should not have the ambition to pursue a profession reserved to the stronger sex. Mme Hatinguais had encouraged us to prepare, for Christmas, a small play at the level of backward teenagers. We had refused. Her replacement, Mme Prenant, was more open-minded. She had been a philosophy professor in Fénelon, appreciated by her pupils and by her colleague, my mother. I had met her and found her likable. I had regretted, when I was in "terminale", not to be able to enjoy her lessons; indeed, under the occupation, she and her husband, Professor Marcel Prenant, had gone into hiding because they were both Communists and Resistants. Mme Prenant, our new Director, invited me kindly to come to her office and asked me what my plans were for the future. She welcomed my wish to do scientific research and advised me to talk with the Vice-Director, Mlle de Schutenbach, a scientist and unmarried former Sévrienne of the class one year after my mother's. When, discussing my future plans with her, I said that I would like to do research in theoretical physics, she advised me to see Frédéric Joliot or André Lichnerowicz. Joliot was a famous physicist, Lichnerowicz was a bright mathematician who was then still quite young. He was the only son of a remarkable couple, a literary and a scientist, like my parents. In his case, it was his father who was Agrégé des lettres, the son of a Polish General Lichnerowicz who had fled his country after having participated in an insurrection; he had an important position at the Alliance Française. The mother, Antoinette Gressin, had been a brilliant student in mathematics in Sèvres. Antoinette had stopped working after the birth of her son, who was the apple of her eye. André was devoted to his mother and was anxious to live up to what she dreamed for him. Mlle de Schuttenbach had been in the same year as Antoinette in Sèvres and had since then remained an intimate friend of hers. She bore, to her and her son, great affection and admiration.

I had not yet made up my choice when it became time to pass the Agrégation.

Despair

The Americans had been long in sinking into Germany and liberating the camps. They were motivated by their wish to spare their soldiers, but it was said at the time that they had made an agreement with the Russians, promising to let them occupy those unfortunate Eastern countries which were liberated by them — and fell under Soviet rule at the Yalta treaty. The death of my father on the first of January 1945 was announced to us during a visit in spring of 1945 by Pauphilet, a literary who returned from London the new director of the ENS. I think that the death of my father was probably a relief for him, confirming him, without possible discussion, in his directorial position. My mother collapsed at learning of the death of her husband. My sister came to see us a little later with her two-month-old baby girl, hoping that the sight of this young life would comfort my mother a little, but my mother threw her only a glance and said, "What do I care?" The eldest of my nieces who has had some psychological problems assigns them, on her psychologist's advice, to these words from her grandmother. I doubt it. In fact, my mother has not been interested in any of her grandchildren, except at the end of her life when the visit of her youngest granddaughter accompanying me to her retirement home seemed to please her. For many years my mother had been angry at the whole world because of her misfortune. I will not speak more of the bitterness that was added to our grief.

My Last Year in Sèvres

My friends and work were support to me. Indeed, the year 1945–1946 was, for me, that of preparing for Agrégation, a contest leading to teaching positions. I had to get to work. Living with my classmates, especially my friends Nicole and Maïtou, was, for me, a great help. We worked on often difficult problems, given at competitions of previous years. We also prepared the lessons, supposed to be given to pupils, that we would deliver in front of the Agrégation's jury. Corrections of problems and critics of lessons were made in classes reserved to Sévriennes. Many of us cut several of them, judging them useless. I must confess that in my case it was fairly often. I came, however, to like one of our professors who embellished his classes with personal comments, Luc Gauthier. He became, later, my colleague and friend.

The competition does not leave me very clear memories. I had worked seriously, often with my friends Nicole and Maïtou, sometimes with Léonce. We searched up the solutions of problems published in Annals of previous competitions, and we looked in appropriate books to prepare oral lessons. The courageous Maïtou went, more often than Nicole and me, to the lessons given by our teachers and communicated their content to us. At the exam, the problem on mechanics was not difficult. I was surprised that two of the questions were identical, except for the numerical application, but I wisely did not comment on that. The problem of mathematics was more interesting. The last question was difficult. I wrote a solution I was not sure of. After the competition I asked Gauthier who had given the problem if my solution was correct, he answered, "Perhaps, I did not find it myself!" After the competition, we had an interview with the jury's president, a General Inspector of National Education, and a member of the jury, a university professor. In my case it was Lichnerowicz who had replaced, in the oral part of the competition, Henri Villat, who was too old and deaf for the job. The inspector told me, "Your lesson was good, but the level too high for pupils." Lichnerowiz told me, "Your lesson was good, but you could have made it of a slightly higher level." The inspector gave me some pedagogical advice; Lichnerowicz asked me what I would like to do in the future. I answered, "I would like to do a thesis on theoretical physics but I don't know if I am able to do it." He said: "You will do one if you do as I say. I work on General Relativity and I am interested in philosophy." I said, "it suits me very well." My professional fate was thus set. I never regretted it, and I remain grateful to Lichnerowicz. In those days there were few women doing research, but Lichnerowicz had no prejudice against women, perhaps because of his attachment to his mother. After me, he had many other women students who became university professors, an unusual fact at the time. In general, he gave women subjects concerning General Relativity, to men he considered the most gifted, he gave subjects of differential geometry. The subjects Lichnerowicz proposed led to some immediate results and he was always encouraging his students with compliments on the results they had obtained. Thanks to this encounter, I had the good luck to do what I dreamed of in my youth: to try to know this universe in which we live. I have not made the astronomical computations which my father feared, those are now the computer's undertaking. I have done, however, rather a lot of calculations, algebraic ones leading logically to mathematical formulations of general physical laws. One may hope they are followed by the reality in which we live. It is exciting work.

My friend, Georges Charpak, chose the other part of the alternative quoted by Mlle de Schuttenbach: he entered the Joliot laboratory. His remarkably imaginative mind has led him to construct new instruments to explore this reality. I would have been absolutely incapable of that.

Respite and End

In the beginning of the summer in 1946, the ENS organized a kind of voluntary sports training of about ten days near Saint Malo, for students. I went, and I have good memories of that stay, which my friends, Nicole and Maïtou, also attended. Some likable Normaliens were also participating, among them, Henri Cabannes, my contemporary and friend, and also, a little younger, Jean Pierre Serre, an exceptionally gifted mathematician. Jean Pierre has been the first to receive the Abel prize, the Nobel prize of mathematicians, after having been elected at the "Collège de France" at less than thirty years old. When I congratulated him for that he answered, "It killed Lebesgue." To my surprised remark — "But Lebesgue lived very old!" — he replied with some contempt for my obtuseness, "It killed him mathematically."

Jean Pierre had the kindness to come to Bures sur Yvette to award me the "Grand Croix de la Légion d'Honneur" at the ceremony organized at this occasion by the IHES in 2015. I could then tell him how happy I was that he had not followed Lebesgue's example. In 1946, Jean Pierre was already engaged to the Sévrienne of his class, Josiane Heulot, who became his wife and had been with me in Taupe. She preferred chemistry to mathematics, of which she became a specialist. Josiane has worked with energy and success for the fusion of the feminine and masculine ENS. It certainly has its good sides, but has resulted in the reduction of the number of admissions of women mathematicians. My classmates regretted the disappearance of the warm atmosphere of solidarity which prevailed in our time when their daughters entered in the merged ENS.

I forgot, for a moment, the pain that overwhelmed my family. Unfortunately, when the stay in Brittany ended, misfortune overwhelmed me again. My mother and I went for a long stay to the country house my sister's parents-in-law owned in Puygros, a village at the foot of the Alps. Mme Dalmeida, already a widow, and her daughter, my mother's age, came to stay near us to comfort my mother. I hoped for the company of my brother Francois, but my mother had accepted, for him, an invitation in Sweden, from an aid organization helping war victims. She had refused, for me, an invitation of my uncle to accompany him in Corsica in the family of Tantine, who had liked me, saying she did not want to remain alone. Those holidays were, for me, very hard. At night I shared my mother's room; she prevented me from sleeping and made her grief weigh on me without realizing I was also traumatized. By day I had strolls with my mother and the old Mme Dalmeida, who stopped every moment to recover her breath.

In principle, after Agrégation we had to take a position in secondary education for ten years. In the spring of 1946, Mlle De Schuttenbach offered for me to be an assistant for three years (it was called "Agrégée répétitrice"), a newly created position which would enable me to work on my thesis while teaching advanced mathematics to students of classes following mine for a few hours a week. It was a favor which I accepted, following the advice of my mother and Léonce.

As for myself, I was rather hesitant because I remembered how little enthusiasm we had following these supplementary lessons. I soon regretted having accepted, because the Ministry created, for the mathematicians whose normal curriculum was three years, an optional fourth year where they would freely continue their culture, in the same material conditions as in the previous years. Parisian housing that we occupied were recovered by their American owners, and the Ministry found, for the Sévriennes, prebuilt pavilions on boulevard Jourdan, near the "porte d'Orléans". Nicole and Maïtou did a fourth year housed in these pavilions, as well as the physicists whose studies lasted normally four years. I asked to be also housed at boulevard Jourdan, but was told it was impossible since I was no longer a student. My offer to pay did not flex the authorities. I had to give up living with girls of my age. I went to live with my mother and crossed difficult times.

7

LIFE IN MONTAIGNE

Moving

I spoke, in chapter 4, of how my mother came to live in the lycée Montaigne at rue Auguste Comte, after she had to leave our apartment at rue d'Ulm to make room for the new Vice-Director. The official apartment given by the Ministry to that Headmistress of the boarding of students of higher classes in the lycée Fénelon was located on the second and last floor. It was vast, even more so than the one we had at rue d'Ulm. There was a lounge, a dining room, a large kitchen with a pantry, a bedroom for each member of the family, and even another room which was used as a living room or a guest room. Two of the bedrooms overlooked the Luxembourg garden and were occupied by my mother and my grandmother. At the end of the holidays, being deprived of the right to share the lodging of my classmates and friends, I went to live with my mother in the lycée Montaigne together with my brother and my grandmother.

First Year at Montaigne

François was a brilliant hypo-taupe student. One of his former classmates, Jacques Blamont, who became a famous astrophysicist member of the Académie des Sciences and author of several remarkable books published by Odile Jacob, told me recently that his classmates had no doubt that François would be top of the list for the entrance examination to the ENS. My mother, however, was worried for him, and he was, more than ever, the main focus of her concern.

I had been, up to then, a strong young girl, a little round without being obese. I had lost, without damage, two or three kilos after my father's death, but in September 1946 my health began to decline. I apparently healed without problem from a bronchi-pneumonia but probably then began the primary tuberculosis which took nearly three years to heal, and was diagnosed in view of its consequences by a radiography only about ten years later. Life was very hard for me: the classes I gave did not much interest me, and I lived in fear of upsetting my mother, always ready to relieve her own grief by a reproach of others. My grandmother, now a very old lady, was not immune to this chronic dissatisfaction. My mother seemed to reproach her for being alive while her husband was dead. Mamé found some comfort in her faith, which caused the wrath of my mother, "If you believe in an all-powerful god, why didn't you ask him to save your son-in-law?" I pitied my grandmother but did not dare intervene. However, she felt my sympathy and gave me, when we were alone, the only jewel in her possession, a bracelet in red gold. It has disappeared now, probably in a burglary. I escaped from the house by visiting Léonce, who had rented a room in an apartment fairly close to the lycée Montaigne. I remember reading, there, volumes of the Labiche plays belonging to the landlady, while Léonce was doing mathematics. I also did some when I was not too tired, an escape that supported me throughout my life. I studied the remarkably precise and clear books "Calcul Tensoriel" and "Problèmes Globaux en Mécanique

Relativiste" by Lichnerowicz, who had encouraged me to learn how to compute with tensors. He said that if I became an expert in tensor calculus, I would be the only one besides him (it was in 1946 or 47).

During the Christmas 1946 holidays, our friend, Georges Charpak, proposed that we enjoy, like him, a stay in the Swiss Alps to be organized by an association for help to the war victims. Georges had been admitted at the end of his hypotaupe to the "École des Mines". Liberated from the camps, he had not felt like going back to a lycée to prepare for a more prestigious school, but he had kept, with Emile and Maïtou, a friendship which extended to Léonce, Nicole and me. Perhaps, also, he felt more in tune with us, students at ENS, than with future engineers. We saw him regularly, always with pleasure. At the time, like many young people before the invasion of Hungary by Russian tanks, Georges was still a staunch Communist. One day when I was walking next to him, we met a friend of his, a party member, who asked Georges, speaking of me, "Is she a comrade?" Georges gave the negative and kind answer, "She is a good comrade." That winter, the weather, execrable, convinced us young people to leave the mountains for Lausanne. A long, uncomfortable trip by train took us there, in the company of the two young sisters of the doctor who became, years later, my colleague and friend in the French Académie des Sciences, Emile Etienne Baulieu. It was nice, but I came back from those holidays as tired as when I had left Paris.

Marriage

The holidays of summer 1946 had left me such a bad memory that I did not feel strong enough to renew them. However, I did not have the courage to tell that to my mother. This prompted me to hasten my marriage with Léonce, who had not really wished it, being more in a hurry to go mountain-climbing in Chamonix than to marry. However, he accepted the wedding, approved by his parents, in the town hall, then in church where I entered holding my uncle's arm. Léonce and I took the train for Chamonix that same evening. We had first planned to camp there, but my mother-in-law, a good woman as I experienced later, insisted that we go to a hotel and gave us the money for it. So we went to a hotel, "Le Faucigny". I keep a pleasant memory of this stay. Léonce forsook me sometimes for a climb too difficult for me, but he asked forgiveness for his absences with the gift of a ring that had a pretty aquamarine. After our stay in Chamonix, we went to end the holidays in the small village in Ariège five kilometers from Pamiers, Bénagues, then fairly isolated in the countryside. My parents-in-law spent their holidays there, in a house which the father of my mother-in-law had built with his own hands. My mother-in-law was very attached to Bénagues and her house. The comfort was rustic: no water, no gas and the toilets a small hut in the garden — a luxury, I was told, recently added. This rudimentary comfort did not bother me. I had known similar conditions a few years before in Dammartin. Léonce took care of bringing jugs of water. In Dammartin we had a stove, but my mother-in-law cooked on the hearth. The soups she cooked on the rack had, to her, an irreplaceable flavor. I had nothing against her soups. I liked less her overcooked pasta and steaks, but I appreciated not to have to cook, and later I was happy that my cooking for my husband or father-in-law gained merit from comparison with hers.

Léonce's parents have enlarged this house and improved its comforts when they came to live in it upon their retirement. Léonce kept it after their death, but did not return to Bénagues, liking Chamonix better. His youngest granddaughter contemplates settling there. With her companion, she works actively to put the house and garden back in good condition. They really need it; it is hard work. I think that the soul of her grandmother, if there is a soul, is happy at this rebirth of the house built by her father.

At the beginning of term in 1947, Léonce and I returned to Paris and settled in the lycée Montaigne: I did not have the heart to refuse my mother to continue staying with her, because she seemed very distressed by my possible refusal of cohabitation. My bedroom was of respectable size. My mother bought for herself a single bed, which left more space in hers, and gave us her marital bed and the dining room as an office. Our meals, in the absence of guests, were taken in the pantry. Léonce accepted easily that installation: he had sympathy for his mother-in-law and no rent was arranging our finances. In fact, I think that he suffered less than me from the cohabitation. Besides, I don't know if I would have been better in an apartment alone with Léonce. Indeed, during the years 1948 and 1949, I was extremely tired, losing 5 or 6 kilos. Giving my classes, though few, completely exhausted me. They were held at boulevard Jourdan where the ENS had moved after the return of the legitimate American owners of the Parisian buildings that had housed the Sévriennes during the war. Léonce sympathized, brought me breakfast in bed and called me "pôtiti" for "pauvre petite" (poor little one), perhaps happy of his superiority. The doctor of the boarding, directed by my mother, came to see me and gave me a skin test which proved negative, from which she concluded that I was not having a primary TB. In fact she was wrong; a radiography revealed, a few years later, characteristic lesions explaining my weight loss and tiredness of that time. The Sèvres librarian, who knew me before and saw me then, told me, "If you were my daughter I would send you to spend six months in the mountains." That nice librarian, Mme Fèvre, was very competent and far-sighted even outside of her job; but my mother, obsessed with her own misfortune, made me responsible for my fatigue, advising me to get up early and take exercise. I did not have the strength. An escape was given to me by my research work which started under the benevolent direction of André Lichnerowicz.

First Works, the 3+1 Formulation

Lichnerowicz used to propose, to his students, subjects leading fairly rapidly to some primary results and he encouraged them with compliments. When I started working with him he gave me to read his seventy-eight-page book, "Problèmes Globaux en Mécanique Relativiste", published in 1939 in the series "Exposés de Géométrie" directed by Élie Cartan. It was, in fact, his thesis, directed by Georges Darmois. Out of the ENS in 1936, Lichnerowicz had asked Élie Cartan for a subject for a thesis on geometry. Cartan had proposed a problem on which himself had worked without success. Lichnerowicz has solved the problem, but only thirty years later. In 1937, he had made no progress when he met Darmois in Institut Henri Poincaré and confessed his fear of never succeeding to do a thesis. Darmois assured that General Relativity offered geometrical problems ready to be solved and that if he worked on them he would quickly complete a thesis. Lichnerowicz followed Darmois' advice, read his short, dense and clear book "Les Equations de la Gravitation Einsteinienne" and in less than two years completed his thesis, the beginning of the intrinsic formulation of the Einstein equations and the study of global problems. The book by Lichnerowicz in 1939 contained an important theorem: by using Cartan's exterior differential calculus it shows that an empty and stationary Einsteinian space-time is identical to the Minkowski space-time of Special Relativity (under a hypothesis of behavior at infinity in space). This demonstration earned Lichnerowicz a letter of congratulations by Einstein who had conjectured the theorem and only succeeded in proving it in a restricted case, in spite of Pauli's collaboration. At the end of his life, Lichnerowicz said he had been a student of Elie Cartan. There was truth in that, but I found that it still showed a lack of gratitude towards Darmois.

I studied, without too much difficulty, the book of Lichnerowicz and the Riemannian geometry in Elie Cartan. I could then begin research work, an exciting occupation which makes one forget the hassles of the outside world. I published, in 1948, my first paper at the Académie des Sciences, "Théorème de Gauss en Relativité Générale", a fairly easy solution to an open problem pointed out to me by Lichnerowicz. As was the custom at the time, I brought my paper to a competent Academician for him to present it. It was Elie Cartan, then about seventy-five. He presented my paper after asking me, "Did Lichnerowicz read it?", adding, "If I was younger I would read it myself." I published, in 1948, three other papers in the Comptes Rendus of the Académie des Sciences — among them, two are cosigned par Lichnerowicz. In one of them, we simplify a hypothesis made in constructing the Schwarzschild metric, fundamental in Einsteinian gravitation. Lichnerowicz then proposed to me, as a subject for my thesis, the generalization of one of his theorems, itself a generalization of a fundamental theorem from Georges Darmois. The problem was to formulate the Einstein equations of General Relativity, in replacing the particular coordinates chosen by him generalizing the particular ones chosen by Darmois, by fully general ones. This formulation, called now three plus one, highlights the local splitting of the roles of space and time. I had essentially finished this work, which required little imagination, and published a summary in a paper to the Académie, when Leray suggested to me a more interesting subject. I will come back to that later.

Lichnerowicz was elected in the Sorbonne in 1949. Before that he was a professor at Strasbourg University. However, he came often to Paris to see his mother, a widow. An only son, he was much attached to her. I saw Lichnerowicz at his mother's, she lived like us near the Luxembourg garden. I also went a few times to see him in Strasbourg, lodging at the Rectorat at my uncle's. I keep a pleasant memory from these stays. Tantine welcomed me nicely. Breakfast was brought to my bedroom. Tonton, still in good health, was always interesting and affectionate. Lichnerowicz received me at his home. We discussed my work but also other things, always stimulating conversations. I am very grateful to Lichnerowicz for the encouragements he had lavished on me in my early days in scientific research — scarcely open to women at that time — and the friendship he had showed me all his life. I was — with Yves Thiry, my elder by ten years — his first student, and though he had many more later, I always kept a privileged status. Long before I was elected there, Lichnerowicz told me, "You will be the first woman in the Académie." I think that Lichnerowicz has, for his part, gained some benefits from our encounter. My works have brought him essential elements for some of his own on relativistic fluids or coupling of gravitational and electromagnetic shocks. Having me in the position of professor at the university P. and M. Curie has given, to our research group, lodging and secretaries.

Holidays

The holidays of summer 1948 were a pleasant diversion. I wished to travel. Léonce accepted that we go to Italy after the obligatory stay in July for mountaineering in Chamonix, under the condition that we include a sports activity, bicycling. I rallied willingly to this project. We bought two bicycles, state-of-the-art for the time, and camping equipment to carry, with Léonce taking the heaviest part. We put our bikes on the train to Genoa and left towards Rome across the Apennines: the coastal highway did not exist yet. I keep some good memories from this hiking: the meetings with Italian countrymen who believed we were in pilgrimage, the camping on the shores of the Trevi lake in a beautiful landscape, the bike rides in the parks around Rome, Adriana villa, the Tivoli gardens, the property of the Pope at Castel Gandolfo. In Rome itself we found accommodation in a student's house. We admired, together, the Colosseum, the forum, the Palatine hill and other wonders of the city, less invaded by tourists than it is now. That year, time and our strength did not allow us to go further south. The road we followed was full of charm, but also of hard climbs and steep descents, despite the inhabitants assuring us that the next leg was as flat as the gesture they made with their hand. We came back home by train.

The following summer in 1949, we gave up on biking and took a train for Naples, then various coaches, to go further south. These first visits to Paestum, the Amalfitan coast, Capri, have left me with enchanted memories, which I have not forgotten in spite of renewed stays in that wonderful region of Italy.

Before going to Italy, we had had a pleasant stay camping in Collioure with our friends Pallarès, Emile and Nicole born Fontanel, and also Georges Charpak and his parents. I had a great friendship with Georges and I sympathized with his mother, a woman not much more than forty, open and courageous, who spoke to me of her youth in a region which was then Polish and now Ukrainian. It was very cold in winter, but the isba where she lived with her family was so well-heated with a wood stove that one could go outside for a while without covering up more. She had married — very young — an older man and soon had her two sons. Her motherly love had excluded, for her, any wish to change companion. In 1931, because of anti-Semitic persecutions, the family had exiled itself in France. Mme Charpak, intelligent and courageous, had quickly learned French and started to work to support her family, helped by her husband. She was very anxious to give her sons an education which would enable them to integrate themselves perfectly into French society. She succeeded remarkably with the eldest, Georges, the inventor of the wire chamber used for detecting elementary particles, and Nobel Prize winner, who gave his name to numerous streets and schools across his country of adoption. Georges' three children are listed on the Internet for various accomplishments. Their grandmother would have been proud of them. When we chatted amiably in the big tent that the Charpak had rented for their holidays in Collioure, these successes were yet to come, but Mme Charpak had been full of confidence. I think she found me likable and would gladly have welcomed me as her daughter-in-law but I was already married. Incidentally, at that time, Georges had no desire to tie the knot. We had a happy stay in Collioure during which Léonce and I, sometimes with Georges, took long bicycle rides climbing the Pyrenees passes, as Georges relates in his book "La Vie à Fil Tendu".

I Become a Mother

I had been an "Agrégée préparatrice" (teaching assistant) in Sèvres for three years, required to give classes that weighed that much more on me that I judged them useless. Meanwhile my male comrades, researcher apprentices like me, were research assistants at the CNRS without being required to teach. Our Science Director, Mlle de Schuttenbach, as a favor, offered to me a fourth year in my position, but upon asking advice from Lichnerowicz, I refused and I too became a research assistant after the summer of 1949.

Léonce and I, then both free of present requirements, forsook the Christmas holidays of 1949 to stay with my grandmother. This allowed my mother to take a holiday away from Paris with my brother. Our positions as true mathematics researchers allowed us to go skiing in the Alps, at Notre Dame de Bellecombe, for a full month in February. We stayed in a modest but comfortable hotel. I was already winning against the tuberculosis bacillus, but the time in the mountains, free of worry, completed my recovery, as Mme Fèvre had foreseen.

Léonce had told me, before marrying, that he did not wish to have children, but I had heard other young men say that and change their minds later. In any case, if a child happened, they became admiring and affectionate fathers. I myself had always considered motherhood a natural and happy part of a woman's life. We were almost three years married, and I was now in good physical condition. I wanted a baby. Léonce's care to avoid it loosened one day and I immediately became pregnant. Léonce resigned himself to this, but with regret. His mother however, deeply rejoiced, having always regretted having an only son. She convinced Léonce to give up plans for Chamonix that summer, replacing the traditional Alpine mountaineering with a hotel stay in Ax les Thermes, a tourist village in the Pyrenees above Foix, not so far from Bénagues where we spent part of the summer holidays every year. That stay was, for me, pleasant and restful. I befriended another guest, a little older than us, who knitted a vest for the baby-to-be. Léonce worked on his thesis. Almost every day, we took a walk on some of the surrounding trails that present magnificent views. I was a little saddened when some of Léonce's mountaineering friends took him for a climb higher in the Pyrenees, because upon coming back he told me, "At last I had a few pleasant days."

Life went on after our return to Paris at the end of September. On the first of November 1950, contractions announced the beginning of delivery at the due date. Léonce took me to the clinic and followed me into the room I was given. His compassion towards me felt rather cumbersome to me, and I suggested he go take a walk to relax, which he did. I went without him in the birthing room, which incidentally was the custom at the time. Childbirth is never a pleasure cruise. Mine was not particularly difficult, but the doctor insisted anyway to make me breathe in some chloroform while he performed an episiotomy (sliced me). When I came to, my daughter was born, during the night between the first and the second of November. I was presented with her already swaddled. She was a pretty, fresh and pink baby weighing 3 kg 150 g. I stroked her cheek to welcome her. Léonce showed relief at the outcome of the birth, but no pleasure at seeing his daughter, however cute she was. My mother did not bother to come visit her granddaughter, whom I named Michelle. I had hoped for a boy, imagining that my mother would bond more easily with him than with a girl, and had chosen Michel as a first name. I myself had no regrets about Michel being Michelle. I even think that my mother would not have been more interested in a boy she did not make herself. She told me once that one loves their children because they are one's creation. She also told me later that if she did not come to see us, her newborn granddaughter and me, at the clinic, it was because Léonce had told her to his satisfaction that I was no more interested in our daughter than he was, which had shocked her! Not having witnessed the conversation, I cannot attest to that account, but I think it was a pretty bad excuse for my mother.

I wanted to breastfeed my baby and I started to put her at the breast, but I was ignorant and the staff of the clinic were unwilling. I quickly had painful cracks. As I was insisting to continue breastfeeding, somebody set a breast pump and a feeding bottle on my bedside table, with no explanation. I did my best, but only managed to pump a small volume of milk that gained me a sneer from the midwife who came into my room to show me a very fat baby, saying, "This is a fine baby." I did not share her opinion, but I accepted giving up breastfeeding.

The doctor who helped me deliver the baby insisted that I stay abed for a fortnight, at the clinic. It was the fashion then. I was not tempted by that stay and Léonce strictly opposed it. Thus I went home to rue Auguste Comte, staying abed as much as possible. Léonce insisted to take care of most of the mundane tasks, but never manifested any pleasure in the existence of his little girl, a beautiful baby. Luckily, my master Lichnerowicz had a pediatrician, Dr. Lepintre, whom he trusted completely with his children. That doctor lived near the Luxembourg, on rue Dugay Trouin, not very far from us. He readily accepted to take care of Michelle. He was a very competent doctor, married to a hospital neurologist. He also was an intelligent and kindly person with whom I became friends. Dr. Lepintre came all the way to my house to examine my children when I lived in the suburbs. He supported me during the youth of my three children, because I had high regard for his diagnosis and counsel.

Thesis

In 1947, before our wedding, Léonce and I had followed an advanced course from a great mathematician, Jean Leray, who had become a professor at the Sorbonne after his return from five years of captivity in Germany. Leray, born in 1906 and dead in 1998, was one of the greatest mathematicians of his generation. I will say more about him later. He was a pillar in my life, a master and friend who supported me for more than fifty years.

Leray's course in 1947 did not lead to any diploma. He had a small audience, and asked for our names. When he heard that I was the daughter of Georges Bruhat, who he knew had died in deportation, he immediately showed sympathy. Upon learning that I was preparing a thesis under the direction of Lichnerowicz, he read with attention the latter's book "Problèmes Globaux en Mécanique Relativiste", and advised me to abandon the subject given to me by Lichné for a more interesting one noted as open in that book: the resolution of Cauchy's problem for Einstein's equations, in the non-analytic case. Analytic functions are determined in their entire domain by their value in the neighborhood of a point, thus they cannot detect propagation phenomenon, nor causality. It is a very important physical property that Newtonian mechanics does not possess. Newton himself was bothered by the fact that in his theory, the action of gravitation propagates with infinite speed. I relayed Leray's advice to Lichnerowicz who answered, "It is too difficult a subject for a beginner." In fact, an in-depth study of general systems of partial differential equations was underway: their classification and the different properties of their solution were booming. It had begun a short time before the war, started by Leray himself with the collaboration of a Polish Jew a little older than him, Julius Schauder. That mathematician of great worth had not wanted to follow Leray's advice to leave Germany when there was still time and had fallen victim to the Nazi slaughter. Leray advised me to read Schauder's articles to try to apply their results to the equations of Einstein's General Relativity. Those articles only concerned linear equations, and for the best of them, equations with only two variables. I set myself to the study of that article, not without trouble, as it was written in German, a language I had never studied. Luckily, by chance, I stumbled upon an article in French, written before the war by a great Russian mathematician, Serge Sobolev, who built non-analytic solutions for linear partial differential equations of the type taken by the linearized Einstein equations in some privileged coordinates called "harmonic", or "wave". Sobolev's method, unlike Schauder's, is constructive and allows for an explicit calculation of the solutions. In his article, Sobolev introduced the generalized functions, now often called Schwartz distributions from the name of the mathematician Laurent Schwartz, who had generalized them some more and studied their properties a lot. Sobolev used them to prove that the functions he was building were indeed solutions to the equations considered. I realized that in fact his proof was erroneous: it only showed the uniqueness of the solutions, not their existence. As for me, I was glad to trade the study of Schauder's articles for the study and completion of Sobolev's, which was not very difficult. I succeeded in using it to solve the non-linear equations of the type of Einstein equations in harmonic coordinates. I also demonstrated that a solution of these equations satisfy the original geometric equations if, at an initial instant, some equations called constraints are satisfied by the data. This last result had already been obtained by Georges Darmois in the analytic case, but it was remarkable for it to still be true in the incomparably more general and physically interesting frame of what became my Doctorate thesis.

Sometime after Michelle's birth, I finished writing my thesis. I can still see myself pushing her carriage at the Luxembourg, sitting in an armchair when she was asleep, rereading my writing or studying the article that I had to present for my second thesis as was the law at the time. Lichnerowicz had modified the subject at the penultimate minute, so I would present some of his own recently-completed work. I was, however, ready on the day I had to defend the thesis. My jury included my thesis director Lichnerowicz — now a professor at the Sorbonne — and Jean Leray. The president of the jury was the dean Joseph Pérès. Marcel Riesz, a famous mathematician of Hungarian origins, Professor at Stockholm and specialist of partial differential equations that my thesis considered, was passing through Paris, and joined to the jury. The scientific thesis defenses are not followed, like the literary ones, by an aggressive debate, but usually only by a few questions to further the presentation. My thesis was no exception and the jury, after a short deliberation, granted me the title of Docteur ès Sciences with the distinction "very honorable". My mother had come to attend my thesis defense, and congratulated me on my dress and my elocution. She had organized, in the lycée Montaigne, the reception that traditionally follows a thesis defense. She met, there, Jean Leray; they both appreciated the intelligence of their interlocutor.

My thesis was recognized as serious work by the mathematicians and earned me, without delay, the inclusion on the list of suitable candidates for the office of Senior Lecturer (it is now called "Second Class Professor"). At the time, this was granted by a committee of professors chosen by the Ministry of National Education, called "Consultative Committee". I had the chance to see the importance of the results I had obtained grow with the return into favor of the theory of General Relativity, which had at the time been somewhat set aside by the physicists absorbed in Quantum Mechanics. Indeed my thesis gave a rigorous mathematical demonstration of the propagation of gravitation with the speed of light, predicted by Einstein, with the consequence of the existence of gravitational waves that, after almost fifty years of effort were finally recently observed.

Léonce's thesis, directed by Valiron, was "Some Properties of Riemann Surfaces", an abstract and arduous subject. He defended it the same year and also earned the title of Docteur ès Sciences with the "very honorable" distinction. Valiron, unlike Lichnerowicz, was not very pleasant. When I went to tell him a few amiable words at the reception — also given by my mother — after the thesis defense, he told me contemptuously, as if my words were from a beggar to her hoped-for benefactor, "Don't worry, I will support your husband." Léonce did not need it — his very serious work deserved, and obtained, like mine, his inclusion on the list of suitable candidates for the office of Senior Lecturer.

First Congress

The first congress I attended was in Liege, a colloquium on partial differential equations in spring 1951. I was taken there by Leray in his car; he held a lecture there. Mme Leray came along with us. We had lunch on the way in a little restaurant. I keep an excellent memory of that trip. Lichnerovicz gave a lecture on the existence theorem for Einstein's equations in harmonic coordinates that I had demonstrated in my thesis. Lichnerowicz said, "These equations are of the type of Madame Fourès" and a facetious listener commented, "It is a pretty type." Bureau, a Belgian specialist of partial differential equations, pointed out to me, in private, that the solution to these equations did not necessarily satisfy the original Einstein equations. I was pleased to answer that he was right but that I had also demonstrated that this property was satisfied if the initial data verified certain constraints. I later set myself to the resolution of these constraints.

To attend the congress, I had left Michelle in the care of her father and grandmother for a few days. My mother did not concern herself at all with it, worried, as she told me, about my brother who was taking the Agrégation exam (he passed as valedictorian as expected).

As for Léonce, in order for his daughter to drink her bottles more willingly, he had added plenty of sugar to them, leading to a worrisome diarrhea that had me call Dr Lepintre. A simple treatment returned everything to normal, but I did not entrust my daughter to her father anymore. I left her to that grandmother, with a devoted home help, to attend a colloquium in Royaumont ten years later. Granddaughter and grandmother appreciated the episode, but it was an isolated one. Fortunately, Michelle's other grandmother was always glad to look after her if we brought her to her house.

Holidays 1951: Lubac and Chamonix

Léonce was a climbing and mountaineering enthusiast, virtually a professional lead climber. Gustave Choquet, native to the northern plains, knew nothing of the mountains when he arrived in Grenoble as a professor, but he was a robust man who loved nature and physical exercise. Octave Galvani, a fellow ENS mathematician like him in position in Grenoble, was an experienced mountaineer from the area who initiated Gustave into the joys of mountains and climbing. Gustave was an exceptional mathematician whose varied interests sometimes overlapped with those of Léonce, and the two men became friends. Octave was friends and mountaineering partners with Robert Bourgeon, a resident of a little hamlet in the Valgaudemard: Lubac in the Hautes-Alpes. Octave introduced Robert to Gustave. Gustave was then married to his first wife, Marie, who had wished to leave him despite the three children they had had together. However they still lived together in 1951 and, for the summer, had rented a house in a hamlet close to Lubac: Les Barangeards. Gustave did climbs with Robert. That year we had rented, for the month of July, a little house in Chamonix, the mountaineer's paradise. Gustave convinced Léonce to go, for some time before, to Lubac where Robert rented out a few rooms.

When I met him in the summer of 1951, Robert Bourgeon was about thirty-five years old. He was rather handsome, tall and lean, sporty with wavy brown hair. He was, if I remember correctly, the biological son of a high-ranking military officer, and his mother had entrusted him, at a very young age, to a childless couple who adopted him. The adoptive father was a shepherd in Lubac. Robert loved his adoptive father. He kept good memories of the summers when he brought lunch to his father in the alpine pastures where he kept his sheep. When I met Robert his parents had died; he lived alone in the rustic house that had been theirs which he had inherited. On the first floor were a living room and a kitchen with a cooking wood-stove. Next to it was a little storeroom–warehouse. Robert had converted, on the second floor, three rooms that he rented in summer to tourists undemanding of comfort. The house did not have running water or a toilet, but it was surrounded by a vast meadow crossed by a lovely stream and encompassed with nice thickets. Robert rented a piece of that meadow as a campground for groups of scouts.

Robert had done minimal studies, but he was a very intelligent man who had read a lot, whatever fell into his hands. His mind was open and curious about his fellow beings. He was passionately attached to Lubac and the mountains surrounding it. He knew them perfectly, and was a competent mountaineer glad to serve as a guide for ascents lovers, although not certified.

Léonce, Michelle and I spent time at Robert's for part of the summer of 1951, semi-camping but also renting one (or was it two?) room so Michelle, still a baby, would have a safe shelter. During Léonce's absences mountaineering, I often talked, with pleasure, with Robert. He liked children a lot, even babies, and he happened to look after Michelle several times to allow me to take some walks on trails impassable to strollers.

After this short stay in Lubac at Robert's, we went to Chamonix. We had rented a little house, two bedrooms separated by a kitchen-living room. My friend Adeline expressed the desire to visit us there, wishing to make up for her refusal, imposed by her parents, to join me for the summer holidays of 1946. Gustave joined us for a few days to mountaineer with Léonce. Thus we received, simultaneously, Gustave and Adeline, one of the bedrooms being occupied by the women and the other by the men, and with baby Michelle's cradle in the central room. I do not remember the ascents that the men took, nor precisely the walks I took with Adeline and Michelle. After that stay in Chamonix we went, like the previous years, to Léonce's parents' in Bénagues.

The Invitation to America

Jean Leray was a great mathematician. The Institute for Advanced Study (IAS), the most famous research center in mathematics and theoretical physics of the United States, had been created in 1930, particularly to receive Einstein who was going into exile from Germany because of the rise of Nazism. In 1950, this Institute was headed by a well-known physicist, Oppenheimer, a father of the atomic bomb who later did his best to oppose its proliferation and use. This caused him a lot of troubles under McCarthyism. A mathematician of great worth, Marston Morse, who admired Leray a lot and who was a permanent professor at the IAS, arranged for Leray to be offered a permanent position, which gave him the right to choose two assistants who would be remunerated by the Institute. It was a prestigious and tempting position. However, Leray was very attached to his native country, France. He compromised by accepting, for five years, a half-time position, his assistants to be employed for two years. Leray offered the positions as his assistants for the academic years 1951–1952 and 1952–1953 to Léonce and myself. We were very happy to accept. My mother had taken interest in life again; my brother was Agrégé and essentially lived with her. She considered our departure for the New World with no regret, wishing us to find, there, happiness that, she could feel, had not been perfect until then.

8

A NEW LIFE, AMERICA

Travel to Princeton 1951

We were expected in Princeton at the beginning of October but we decided to arrive at the beginning of September, to have time to familiarize ourselves with the place and give some practice to the Shakespearean English we had learned at school for ten years. It was a logistical error because air conditioning did not exist yet and heat in September in New York, as in Princeton, was overpowering.

We had found, to take us to New York, a boat leaving from Barcelona, easier to reach from Bénagues than Le Havre or Cherbourg. I was happy of this opportunity to do some tourism in Spain. We stayed in Barcelona for only two days, pleasant though I did not recognize anything when I went there again in 2011 with my youngest daughter. Our boat, the De Grasse, was a bit dilapidated; it was on one of its last trips, but we were not used to luxury, and we found ourselves quite well in this floating hotel. I was a bit sorry that it sailed the ocean: I was not really seasick, but I did not feel perfectly comfortable. Michelle, aged ten months, seemed quite at ease, circulating briskly on her youpala. Peace reigned between Léonce and I. However, we had sympathized with another young couple, parents of a small boy of Michelle's age, and I was saddened to see the joy they shared in taking care of their son, a joy never manifested by the father of my daughter. He helped me to take care of her as a form of duty, never out of pleasure.

Once in New York, we took the train for Princeton, then the big car of the Institute (called "station wagon") led us to our home. In 1950, the IAS was a beautiful building called Fuld Hall. Inside was a large lounge, where tea was served every afternoon, and a well-furnished library with a competent and amiable young librarian. The few permanent professors and distinguished visitors had great offices. The young researchers with a fellowship also had offices, but smaller.

Permanent professors had their houses; distinguished visitors, for instance Leray, were housed in pretty villas. Young researchers like us were given wooden pavilions built during the war for soldiers, dispersed in a large lawn near Fuld Hall. Ours had a living room with kitchenette downstairs and upstairs two bedrooms, one for us and one for Michelle, and of course a bathroom. We found ourselves very comfortable there as soon as the hot weather had passed.

Daily Life

I was young, and I had regained health and the will to live. A woman doctor recommended by Mrs Morse had found Michelle underfed and changed her diet, prescribing a daily egg. My question, "Won't it be bad for her liver?" had made her laugh. Michelle prospered. She refused, for several months, to walk and to talk but the doctor drove my fears away in a few words, even saying, "I think she is very clever" when the small one waved good-bye when she walked to the door. One day Michelle stood up and began at once to walk confidently. One of her first words was "Lucky", the name of a neighbor's dog that Michelle said while looking at a running ant. This animal identification seemed to me to confirm the doctor's remark.

After an unfortunate trial with the wife of a colleague, we hired a fifteen-year-old girl to look after Michelle in the afternoons when leaving school. Michelle and her babysitter were quite happy together. In the absence of our babysitter, Léonce took over my care of Michelle sometimes. Fortunately, mathematics can be done anywhere, and the presence of our daughter did not hamper my work. I had noticed, passing in the lounge to have tea, a number of my colleagues immersed in that game of Chinese origin called "Go". One of them said to me one day, as I was leaving the lounge to be home before the babysitter left, "Obviously a woman cannot be a mathematician, she must take care of her children." I answered, "She cannot linger to play Go."

Life in Princeton was, for me, a happy time. America was a new world, full of vitality. Material life was very easy, in contrast to the difficulties we had known in France during the war. Human relations were warm. Professors and students treated each other as equals, using their first names, an unthinkable familiarity in France at the time. The permanent professor Marston Morse, a great friend of Leray, welcomed us amiably in spite of his anti-feminist prejudice. After a lecture I had given on my work, he asked Leray, who repeated it to me, "Is all that from her?" His comment after the affirmative answer from Leray was: "There are remarkable works by Emmy Noether, but she was not a woman." It is true that she, of whom I was just gifted a bronze tablet, had a rather severe face. Bochner, a mathematician professor at Princeton university, strong in mathematics but not in politeness, asked me, "Is it Lichnerowicz who wrote your thesis?", adding,

"When a woman publishes an interesting mathematical work, it is a man who did it. Sophie Kovalevski was at her best with Weierstrass." This insinuation led me only to think, "What an unmannerly fellow." I don't know if he was married.

Marston Morse had an intelligent and charming wife, his second spouse, who was much younger than himself. His first wife had caused a scandal by leaving him, after ten years of marriage, for a much older mathematician, taking, with her, their two children. Marston had made, with the second Mrs Morse, five children, like Charlie Chaplin with Oona. This second wife was, like Oona, a remarkable woman devoted to her husband and children and even stepchildren. Mrs Morse did her best to make my life easier. She was a hostess full of kindness for passing foreign visitors that she and her husband took to a restaurant. I remember, with amusement, such a dinner where I had been invited together with a French mathematician, Jean Favard. He was in his fifties, a professor in the Sorbonne, renowned for his scientific rigor and his humanism; he also was a bon vivant. I don't know if prohibition reigned then in New Jersey, but a stylish server filled our glasses with only fresh water. In the middle of the meal Favard's glass was still full. Mrs Morse called the server and said, "Change this water, it is no longer fresh." Favard did not resign himself to empty his glass of renewed water. When going out he grabbed Léonce's arm saying forcefully, "Is there not a place in this town where one can drink a beer?" A young American who was with us knew such a place and took Favard, who could at last quench his thirst. Favard, a friendly man of robust appearance, died comparatively young, at 63, from liver cancer. Respect for Mrs Morse's attentions would perhaps have preserved him from it. Jean Favard was from the province Limousin, and very attached to his origin. A school bears his name in the Limousin town, Guéret.

New Friends

Tea was a pleasant time, almost an obligation, providing, to fellowship holders and professors, an occasion to meet, not only to discuss science, but also to better know each other. In the beginning of my stay in Princeton, in spite of my ten years of academic study of English, I understood nothing and made myself understood with difficulty. I remain grateful to Irving Segal, who patiently came to keep me company and held, with me, a conversation of which I understood little, but was, for me, an excellent lesson in American English. Thanks to him I could fairly quickly communicate with my American entourage in their language. Irving told me one day, because he preferred to be provocative than complimenting, "You have learned English very fast, you will forget it as fast." He was mistaken because he had taught me only the practice of this language which I had already learned at school for ten years. Our scientific interests were close: we were both interested in the mathematical formulation of fundamental physical problems, among which cosmology. I was faithful to Einstein, but Irving had his own model for the universe, based on simple mathematical ideas. He did not believe in the Big Bang, but in an eternal universe where unchangeable geometry is not determined by its energy content and the resulting gravitation, but only by a property of invariance of space-time under the Minkowski group of Special Relativity and the topology of space, the three-dimensional surface of a sphere. Segal comforted his choice with a deep study and an original interpretation of the results of observations by astronomers. I told him one day, "It is an interesting theory", he replied, "It is not interesting, it is correct." His cosmological theory, disputed at the time, seems to have been beaten to death by the observations made these last years by the Planck satellite and big telescopes.

I continue however, to quote among others, Irving Segal's cosmos, by friendly loyalty and also because I think the universe possesses still many mysteries, perhaps surprises, and alternative models deserve to be quoted. The mathematical problems met by Irving and me were near. We worked together on one of them in Princeton, arriving at a small co-signed article, "Causality and Analyticity", while pursuing research on our personal subjects. We cosigned our second article thirty years later on quite another subject, "Global solutions of the Yang-Mills equations on the Einstein cylinder", that is, with another interpretation, the Irving Segal universe.

Through Irving I knew his friend Georges Mackey, then a bachelor, who had come to visit him. Georges told me, and Irving denied, that he was a former student of Irving, though a little older than him. I would see Georges again several times in the passing years as I had befriended this eminent mathematician interested in physics, who became a professor in Harvard. He was a simple and direct man, gentle, intelligent and perfectly honest. Later, I also appreciated the wife he chose, Alice. Like Léonce, he said he did not want children, but after the birth of his daughter Anne he was very fond of her. Being with them in Les Houches, I saw he always reserved time in the evening for a stroll with his little girl.

In Princeton, Léonce had sympathized with Lisl Novak, a specialist of a subject which interested him, mathematical logic, that was too abstract for me. Lisl, the only woman with a fellowship like me at the Institute, was quite likable; of Czech origin, she understood French, though she refused to speak it. She undertook to give us English lessons, correcting our grammatical errors and our vocabulary. I understood at last that each language has its genius and that no thought with a bit of complexity can be perfectly translated from one language to another. An interesting discovery after ten years of studying a foreign language at school! A few weeks after our arrival in Princeton, Léonce and I understood, and spoke fluently, the country's language. We had made other friends among our contemporaries and we were a small group who met, often in our home because Léonce and I were the only couple. There were, among others, in addition to Irving and Lisl, three Hungarians: Horwath, Fary and Gal, and a Belgian, Jacques Tits, a precocious mathematical genius barely over twenty years old, full of life and with varied interests. Tits was awarded, years later, the Wolf prize, then the Abel prize, the Nobel prize of mathematicians. In the fifties, Segal had recognized his gifts and invited him to Chicago where Irving was then a professor. Later, Tits found positions in various famous universities of the US or Germany. He stayed in France at the beginning of the IHES and finally decided to apply for a position in the Collège de France. He obtained it after getting the French nationality, necessary at the time to hold such a position, the most prestigious in French science. I always remained friends with Jacques Tits, especially as he was a privileged collaborator of my brother François, co-author of "The Bruhat-Tits theory". Jacques and I were elected the same year at the French Académie des Sciences, Tits in the mathematics section, I in the section which was then called "mechanics". We then had many occasions to meet, talk on various subjects, and sometimes evoke our common memories. I remember asking him a question on group theory which I knew to be of his specialty, while its modern developments using a new language were, for me, difficult to follow. Jacques took the trouble to write up for me a detailed answer to my question, using terms which I could easily understand. Jacques married a very nice and intelligent historian. They did not have children, but were an extremely united couple, spending most evenings reading together the same book!

During the first part of our stay at the IAS, from September 1951 to June 1952, Léonce and I had made friends with our nearest neighbor Robert Finn, called Bob, of about our age, a specialist of mathematical problems for classical fluids. We had the opportunity to see, without hampering our good relations, the differences between our cultures, mathematical and otherwise. Bob came to have lunch with me when he was passing through Paris years later, and we had a good time evoking pleasant memories from the past. Bob is now Professor Emeritus in Stanford. His wife, Peggy, who was very nice also, did not have a profession when we were neighbors. She knew French and was hired by Leray to translate, to English, his course on partial differential equations which he had written in French. It was German that Leray had learned at school and he had been a war prisoner in Germany for five years. English was, for him, terra incognita. He asked Peggy Finn to write his courses in phonetics. I listened to the first of these courses, which were absolutely incomprehensible of course. Leray was courageous and determined, as well as hard-working and intelligent. He set himself to the study of English, gave up phonetics and could fairly quickly give understandable courses in English, thanks also to the fact that mathematical equations have universal notations. A few months later Leray spoke English as well as we did, never missing an opportunity to practice, even reading books in English in his spare time.

Marguerite Leray

During that stay in Princeton, I had the pleasure to get to know better the wife of Jean Leray, Marguerite. She was a very nice woman, direct and friendly, on leave from the teaching of mathematics in high school. Three years younger than her husband and coming from the same region, both loving the sea, they made a couple united in depth. They loved their three children as good parents. Leray, with his passion for his mathematics, was not always an easy husband, but Mme Leray was outspoken and did not hesitate to complain. She told me they had started their honeymoon travel along the Brittany coast, but, following an idea that had come to Leray for a mathematical theorem, they had stopped in a small hotel and spent, there, a few days with Leray dedicated to the theorem he had guessed. His wife had not appreciated this interruption. Before the war, M. and Mme Leray had a son, Jean-Claude. Mme Leray had brought him up alone during her husband's captivity in Germany, while teaching in high school. Upon his return, Leray made twin girls with his wife. Unfortunately, one of them died in the clinic where they were prematurely born. The second one, Françoise, had then been between life and death. Leray took things in his hands, brought Françoise out of the clinic to his home and summoned a famous pediatrician who saved her life. Assured of the baby's health, Leray went to a mathematical colloquium where he had been expected; his wife bore him some grudge for that. Françoise became a beautiful and robust woman, a biology research director at the CNRS and mother of three boys. Soon after the twins, the Lerays had another son, Denis, who became a pediatrician. When we were in Princeton with baby Michelle, the Lerays were there with Françoise and Denis, who were old enough already to go to kindergarten. They lived in a villa some distance from us. However, our families met regularly for some Sunday walks. Mme Leray has always been very friendly to me and I have always been happy to see her. When I went to her at the crematory ceremony of Leray, in Nantes, she greeted me with these words in a warm tone, "Ah, Yvonne, my husband loved you." I answered, and it was true, "Me too, I loved you both."

The Course of Leray on Hyperbolic Systems

Leray wrote, at Princeton in 1952, in English, the object of his course, a monumental work on general systems of hyperbolic partial differential equations. This work was published as mimeographed papers. It appeared as a book only in its Russian translation by the remarkable mathematician of various skills, Naïm Ibragimov, who was a very friendly man. These papers contain, however, fundamental results on the definition of "Leray hyperbolic systems" and the existence for non-linear equations of local in time solutions. In this work Leray introduces a geometric definition of global hyperbolicity, fundamental for the global in time studies in General Relativity. Unfortunately, Leray refused the publication of these papers as a book in English, because he and a Swedish mathematician, Lars Garding, had engaged in an endless competition to refine the results. Important improvements in the regularity properties of solutions in the case of one equation have been obtained after the hard, long and painful work of a Canadian clergyman, Philippe Dionne, who had come to work on mathematics in France. It had been suggested and closely monitored by Leray. After his thesis, Dionne went to New Zealand as a professor in Auckland University. I had the pleasure of visiting him there in 1978. Dionne has done, in Auckland, interesting mathematical works less strenuous than those imposed by Leray. He seemed happy in that country, so far from his native Canada. The extension of the results obtained by Dionne for one equation to the case of Leray's hyperbolic systems is not difficult, but would be tedious to write in the general case. I often used it for the equations of mathematical physics that have interested me. I repeatedly encouraged Leray to replace his mimeographed papers with a book containing the improvements on which he had made his student Dionne suffer, but he answered that I should do it myself. It was beyond my strength. The wealth of this profound work by Leray remained unexploited in its generality.

Visits to Einstein

I was attending Leray's course. We were leaving the classroom together when we met Einstein walking towards his office. Leray and Einstein knew each other and Leray introduced me to him, saying that I had defended a thesis on General Relativity that had been directed by Lichnerowicz. Leray said also that I was the daughter of the physicist Georges Bruhat, who had died in a Nazi concentration camp. Both facts appealed to Einstein's sympathy. Einstein knew Lichnerowicz's name and had esteem for him. Indeed, Lichnerowicz had proven a conjecture dear to Einstein's heart, namely, that any stationary space-time with asymptotically Euclidean space is the Minkowski space-time if it satisfies the vacuum Einstein equations. Einstein himself, in collaboration with Pauli, had proven the conjecture in the particular case of static space-times. The stationary case is more subtle to prove; Lichnerowicz had made, for his proof, clever use of Elie Cartan geometric formalism. In fact Einstein's conjecture was stronger: he thought that the result should hold for any asymptotically Euclidean complete vacuum Einsteinian space-time. The stronger conjecture has been proven to be wrong in the five-hundred-page work of Christodoulou and Klainerman of 1990 — a gravitational field can be its own source. However, the physical intuition of Einstein had some truth in it: his conjecture has been proven to be true by Shoen and Yau for Einsteinian space-times with zero ADM mass, that is, tending fast enough to flatness at space infinity (faster than a Newtonian potential).

Returning to 1951, after my introduction by Leray, Einstein invited me to enter, with him, his office, we were then at his door. He asked me to explain my work to him. My English in the beginning of my stay in Princeton was not fluent, in spite of my having studied it at school for ten years. Einstein told me to speak French, which he understood, but he would answer me in English because he had lost the practice of French. So, that first time, I summarized, in French, my thesis on Einstein's blackboard. Einstein listened with interest, and congratulated me for the results which gave rigorous proofs of properties he had expected from the gravitational field. When I left, Einstein told me I would be welcome in his office any time I felt like knocking on his door, though the next time I should speak English since now I had occasion to practice it. Einstein was right — I had no difficulty with English in our later interviews. It was too late for me to catch a good accent, but Einstein's accent was not perfect either. Free access to Einstein's office was a great favor. In those years, Einstein did not mingle much with other people at the Institute. He did not come to receptions, nor to the daily traditional tea in the common room. He did not come to seminars. It is true that at that period, the early fifties, General Relativity was rather dormant, and the subjects in the theoretical physics seminar directed by Oppenheimer were all on quantum mechanics. The problem of quantizing the gravitational field was not even mentioned.

I have knocked sometimes on Einstein's door; I regret that I did not do it more often. I have always been kindly welcomed as a young researcher eager to learn from a great scientist, a genius who was, nonetheless, kind and unsophisticated. Einstein looked like a very intelligent old man, quite appropriately dressed in pants and pullover with socks in his shoes, despite the legend. He was smiling with benevolence at the young woman I was then.

In those years, Einstein was working on his last unified theory with his assistant Bruria Kaufman, a pleasant hard-working woman of about my age who shared his office. She left the States after Einstein's death, to live in a kibbutz. Though focused on unified theory, Einstein was still interested in classical General Relativity and its open mathematical problems. I showed him the small book of 1939 by Lichnerowiz, "Problèmes Globaux en Mécanique Relativiste". In this book Lichnerowicz constructs what he thinks to be a counter-example to the conjecture mentioned before. Einstein read the section concerning it and pointed out that the constructed space-time did not obey the asymptotic condition of no flow from infinity of gravitational energy assumed in his conjecture. Another time, Einstein encouraged me to prove that his equations have no periodic solution. Now, Einstein himself had published, a few years before, a small article in some South-American journal, which I had read by chance. In that paper, Einstein said to prove the existence of periodic solutions. That article contained a simple mistake, an error of sign which I had noticed. I said to Einstein with a small smile he quickly deciphered: "But you have proven there are periodic solutions." The great man answered with another smile, "But this article was not quite correct" and me, "I noticed it." An ordinary man would probably have held a grudge towards me. On the contrary, Einstein, after this, showed me greater kindness.

In my later visits Einstein explained to me, on his blackboard, the non-symmetric theory he was then working on, its successes and its difficulties. I brought him the book by Eisenhart "Non-Riemannian Geometry". He thanked me though I don't think this book was of real use to him. I worked a little on this last Einstein theory. The equations, in spite of some nice features, are very complicated. I obtained some calculus results, but they led to no valuable theorem. In fact, I was doubtful of the physical significance of this theory, which had been interpreted at that time as unifying, by addition, the gravitational potential and the electromagnetic field. In the beginning of the nineties, T. Damour and S. Deser worked on the equations of this last unified theory of Einstein, after giving it a new physical interpretation. The complicated calculus and the difficulty in interpreting the results suppressed, for me, possible regrets at not having devoted more effort to the last theory of Einstein. I understood why in 1951 I had preferred to pursue the mathematical work I had started: it was to find, for Einstein's equations in higher dimensional space-times, properties analogous to those I had found for four-dimensional ones. The problem is not trivial, dimension four leads to quite special properties of space-times (it may be why we live in them), but the problem in higher dimensions is still mathematically well-posed, and the results obtained are clear-cut.

In spite of my lack of enthusiasm for his unified theory, Einstein continued to be very kind to me, as an old man to the young woman I was then, but also as a physicist of genius to someone eager, like him, to understand something of the world we live in. He was quite open-minded, he told me once, as I left his office after some discussion on his last theory, "All that is perhaps the wrong direction." However, he endorsed the motto: "It is not necessary to hope to undertake nor to succeed to persevere." Those who knew him at that time know that he did not consider as tragic the successive failures of the varied unified theories he tried to construct, still believing in a complete and causal theory of the universe, the work of a God "who does not play dice". As stated by philosophers, "The search for truth is better than its possession". Einstein did not really expect to succeed but he enjoyed his work. It is said that on his deathbed he asked the nurse for paper and pen to make some calculations.

During my stay in Princeton in 1951 and 1952, I did not see Einstein as much as I could, occupied by my work on another subject and my family duties. I regret not having made, to the great Einstein, more frequent visits despite his always-friendly welcome and his visionary comments, which were very deep but also full of common sense. I had promised myself to repair that mistake at my next visit. Unfortunately when it occurred, Einstein was no more.

Meeting Other Great Men

The IAS Director when we arrived there in the early fifties was Robert Oppenheimer. He was a great physicist, author not only of important work in Quantum Mechanics, but also of the first results on the formation of black holes. His name is mostly known to the public for his efficient direction, during the war, of the Manhattan laboratory that constructed the first atomic bombs. He was a supporter of their use to end the war, but with the regret that the United States had not previously given more warnings to Japan. Like Einstein, Oppenheimer, after the war, was opposed to the development of thermonuclear weapons, which brought him a lot of trouble under McCarthyism in 1953 until his rehabilitation in 1963. Leray had told us after our arrival at the Institute that we should not disturb its director, a very busy man; however the same Leray told us later that it would not be polite not to go salute this director; therefore we went. Oppenheimer, a thin, rather tall man with an expressive face, received us amiably with some encouragement and banal words. However, I had the chance to be able to better appreciate the remarkable personality of Oppenheimer on other occasions. I particularly remember a reception at the Institute where we engaged in, I don't recall how, a conversation on Buddhism. I was not ignorant of this subject, because my mother, a philosophy professor, had often spoken with me of this religion that touched on our metaphysical concerns. Nevertheless, I was full of admiration for the erudition of Oppenheimer and the quality of his speech.

I did not have a chance to meet Gödel, a permanent member of the Institute who became famous for his work on mathematical logic and his construction of a strange model of universe where time goes backwards. Gödel, like Einstein, did not come to the common room, and I did not knock on his door. His work was too abstract for me. I also do not think that his strange personality would have clicked with mine. I was interested later in the works of Paul Cohen on set theory, without understanding them completely. But they taught me, without really surprising me, that mathematics are a colossus with feet of clay, a construction of the human mind based on abstract definitions which can be replaced with others, however less fruitful. This discovery did not shock me. I like philosophy and I willingly believe that the human mind creates the image it has of the real world, very difficult, besides, to define as an absolute. Some mathematicians think that mathematics have their own existence, and that their work is to explore it. I can believe it, but this controversy seems to me of little interest, because it brings us back to the search for a definition of the word "to exist".

Going back to more down-to-earth considerations, I will say that I met Hermann Weyl in a staircase. He was a tall and heavy man, an author of interesting works in geometry and General Relativity, but more a mathematician than a physicist. His book, in German besides, "Raum, Zeit, Materie", seemed obscure to me. We exchanged some polite banalities. On the other hand, I have been happy to meet Von Neumann, my senior by only twenty years, who was a passionate man of exceptional intelligence that shone in his face and words. He was not a kindly old man like Einstein, but a man in the prime of life, with a pessimistic view of humanity. However, he told me he appreciated my work in General Relativity. He pointed out to me that the so-called horizon, boundary of the Schwarzchild space-time that represents the surface of the collapsing spherical mass generating the gravitational field, is not a singular submanifold of the space-time and advised me to study this problem. I did not do it then, the paradigm being that matter cannot be concentrated enough to attain this boundary. I thought of it later and I had obtained some results when I was attracted to another topic. Despite Von Neumann, black holes received their rightful place without me. I cannot quote here the immense work of Von Neumann on various subjects. Von Neumann was then concentrating on his pioneering work, which has now taken a development shaking up science and our daily life: the construction of the first computer. It was in Princeton, a big machine. If I did not invent this memory, I was waiting with a crowd, for the response that this future prophet was going to give for the first time to its creator. He asked something like: "Factorize 10", after some hard work the machine answered, "It is a prime number". Recall for the philistines in mathematics that a number is said to be prime if it is exactly divisible by no other (except of course 1 and itself), and that to factorize a number is to write it as a product of prime numbers, 2 and 5 in the case of 10. An assistant said the machine was very clever, it had thought, "If I perform this work correctly, what will they ask me later!" In fact the machine quickly gave exact results when its master had corrected a small error in maneuver, and it is true that one asks more and more of computers.

I also had the pleasure of having lunch with the great mathematician John Nash, future economic science Nobel Laureate, before the break-out of the symptoms of schizophrenia he had to suffer for many years. He was then a tall young man, sweet and friendly. He intervened to defend a French mathematician that a colleague had considered overrated, thinking this judgment annoyed me. Nash basically said, "He is not so bad."

Personal Work

During this first stay in Princeton, I mostly worked on the extension to higher dimensions of the formulas I had obtained in the particular case of four dimensional space-times. This extension is not trivial because four dimensions imply special properties, which might perhaps explain why we live in such dimensions. Space-times with dimensions greater than four play an important role in current theories looking for a unified representation of gravitation, electromagnetism and the other fundamental interactions, weak and strong. I must confess that at the time, with these last interactions not being known, my motivation was mostly mathematical curiosity fortified by the prospect of the resolution of a new and clearly posed problem. My work was successful and I had the satisfaction of seeing it published, in French, in "Bulletin de la Société Mathématique de France". In 1952, English was not yet almost mandatory for scientific publications as it is today. I gave a talk on this work at a small colloquium organized, in Harvard, by Georges Mackey. I had traveled, as in 1951, in Leray's car, but this time in the company of a Belgian mathematician, Georges Papy, who became a specialist of the mathematics teaching problem. Gustave Choquet was very interested in this subject. He did not share all of Papy's ideas and spoke of them to me years later.

I followed, of course, the remarkable course from Leray on general hyperbolic systems. He was introducing many new and deep ideas which I used often, later. However, I had not given up my interest for physics. I attended the seminar on quantum theories directed by Oppenheimer that Einstein never attended. I did not find inspiration for personal work among the theoretical physicists present at the Institute. One of them showed me a mathematical problem he had come across in his own research. It was neither trivial nor very difficult. After some hours of work I brought him the solution, he said, "But the problem is not there." This answer discouraged me and I returned to my mathematical problems, which were there. The main memory I keep from this physics seminar is a discussion on quantum vacua and the intervention of the great physicist Wigner, "But in vacuum there is nothing, there can be only one vacuum." Wigner, and also Einstein, lived in a time when it was possible to think that the universe, at all scales, obeys the logic constructed by humanity in the course of centuries, inferred from observations at its scale.

Return to France, Strabismus Operation and Holidays of 1952

In June 1952, upon the official closure of the Institute's activities, we decided to return to France for the summer holidays. The motives were many: our return allowed us to get our CNRS salaries for three months, non-negligible for our modest finances (I am told that now, CNRS researchers staying abroad can keep their salary) and to avoid the stifling heat we had known in September in Princeton the previous year. Finally, Léonce would be able to go to his dear Alps and Michelle's grandparents would take care of her while we went to the congress of Austrian mathematicians in Vienna with Léonce's climbing companion, Gustave Choquet. We took a boat, which brought us to Cherbourg faster than the De Grasse had taken us from

Barcelona. I keep a dazzling memory of the train trip through Normandy at the end of the spring of 1952. What a contrast between these small green fields enclosed by flowering hedges and the extended flat lands around Princeton! Unfortunately, this wonderful landscape has since been devastated by intensive cultivation. It seems people have seen the damage to the environment, and even to agriculture, caused by the destruction of hedges and they try to replant them. I am happy of it.

Back in Paris, lodging with my mother in lycée Montaigne, I had to go to the ophthalmologist for a problem with glasses. My mother advised me to go see the son of a friend of hers, my childhood friend Paul Surugue, now married and a father, who had become an ophthalmologist. I followed her advice. Paul examined me and strongly advised me to have my strabismus operated on, saying that this operation was now well-developed. My mother had always advised me against it, fearing it would transform my fault into exotropia as it had happened to the sister of a former student of hers. She thought, also, that I was alright that way, my glasses making my eyes look bigger. But at twenty-eight, back from America I had acquired enough energy to resist maternal advice. We had reserved a stay with Robert Bourgeon, Gustave and his family having again rented, for the summer a house in Les Barengears. My parents-in-law, devoted and happy to see their granddaughter, willingly accepted to come and replace me with Michelle in Lubac. My father-in-law said amiably that I was right to do what I could to improve my eyesight. I was, without waiting more than necessary, operated on by Paul. He did it for free and came back the following morning to check on his work. He added a last little touch to assure its complete success. He sent me, thereafter, to a very nice middle-aged lady, who had me do rehabilitation exercises. I did not acquire binocular vision, my brain was not young enough to adapt, but my aesthetic was transformed. The orthoptist advised me not to put glasses on when I did not really need them, which I did for some years. I then saw the male sex, like flies before a honey pot, show me new interest.

A few days after my operation, I went to Lubac, finding Léonce, my parents-in-law and Michelle, a charming little twenty-month-old girl, sitting on a bench in front of Robert Bourgeon's house. When she saw me, her face brightened and she patted the bench to make me sit next to her. It is a nice memory. My parents-in-law returned to their home and life resumed its course: climbs for Léonce and Gustave, small walks for Michelle and me, also, chatter with Robert and his current guests. Among them was Simone, the chieftain of a group of scouts camping in the meadow. Simone was nice, younger than me by about ten years, therefore younger than Robert by about twenty. Robert fell in love with her and courted her, as he had with other girls before. But he was able to appreciate the realistic, determined character of Simone, so full of common sense. He decided to marry her, an excellent thing for himself and for Lubac. Thanks to the determination and help of his wife, Robert obtained a loan from the government to transform his house into a touristic hotel. In fact he kept, almost intact, the original part of the house inherited from his parents, as he was very attached to it despite its primitive lack of comfort. Simone and he had built next to it a modern hotel, which they called "Le ban de l'ours". Simone took a cooking course. With Robert as a handyman, they have made many tourists happy, but I preempt. Let us go back to my holidays of the summer 1952. In September, Gustave was invited to a congress in Vienna which interested Léonce. As for me, I did not know Austria and I was happy to go there. Léonce offered to drive Gustave to Vienna in our car. Gustave agreed. My parents-in-law were happy to watch over Michelle. Léonce, Gustave and I made a pleasant trip, with a short stop on the way to visit Salzburg. Léonce was full of admiration for Gustave and treated him with much more amiability than he treated me. This behavior incited me to try to awake the sympathy of the object of this amiability, hoping some of it would redound on me. In fact, during this trip Gustave paid me more attention than he had done before. Among other things he buttered my toasts at breakfast. Perhaps this new favor was due to my new lack of spectacles, who knows? A few years before, I had passed two boys in a street of Dammartin and I had heard one say to the other, "She would be cute without her spectacles."

After the Vienna congress we went back to Paris, other mathematicians traveling with us. Michelle having recuperated, we took again a transatlantic to New York, this time from Le Havre. The trip was quick and does not leave me precise memories.

Back in Princeton

Back in Princeton at the end of September, we found our bungalow again, but the neighbors had changed. The Finns were no longer there, replaced by other mathematicians, a very nice couple whose children had already left the family home: Martin, a confirmed mathematician, and his wife. She offered several times to take care of Michelle in my absence. My daughter was now going to kindergarten and seemed to like it. The other visitors of the Institute had changed also. A little farther was lodged Isidore Singer, who became a famous mathematician. I had, with him, good relations, and I was a bit saddened when some ten years later he gave an unfavorable report for the publication of the book, "Analysis Manifolds and Physics" that

I had written in collaboration with Cécile DeWitt and which she had sent him. Fortunately, Cécile is not a woman to be discouraged by an obstacle. She sent our book to Elsevier who accepted it at once.

I missed my friend, Irving Segal, who had gone back to Chicago. However, Léonce and I made a new friend, who had just been nominated for the position of permanent professor at the Institute. Hassler Whitney had been nearly forty-five but did not look it. I first met Hassler at a reception at the Institute. He was apparently attracted to the pale yellow of a pretty muslin dress which I had bought, for a nonetheless moderate price, on the Champs Elysées. In fact, I heard Hassler make this confidence to our neighbor as I was passing between our two pavilions, the window of hers being open. Due to my dress or not, we started chatting pleasantly. Hassler spoke very good French. Lichnerowicz came near us for a moment, speaking in English. Hassler told me, "I don't understand Lichnerowicz's French", I answered, "He speaks English." Discouraged, Lichnerowicz moved away and I continued the conversation. When Hassler learned that Léonce was a true alpinist, a lead climber, he was enthusiastic. He was a fanatic alpinist. Interrogated about his career a few years before his death, he cited, as his master, the mathematician Alexander, another mountaineer, while his research director had been the very famous mathematician G. D. Birkhoff. Whitney had made many ascensions in the Alps with the Swiss mathematician De Rham from Lausanne, and he was happy to meet Léonce, another fanatic of mountaineering. Hassler was divorced then, and feminine company was a welcome distraction, as well as the prospect of ascensions in the French Alps with a mathematician and climbing leader. Hassler told me that his wife had left him after giving him three children he loved, because she had been unhappy with her marriage. He did not know why. She had taken him to their garden and made him sit on a bench next to her to explain. She had spoken long without him understanding anything. She had then shouted, "You don't listen to me and I believe you even fall asleep!" I had fun chatting with Hassler and I accepted, willingly, his invitations to the skating rink or to a social dinner, but I completely agreed with the affirmation he made one day to Léonce, "I will not steal Yvonne."

Hassler Whitney visited us the following summer in the apartment we had rented in Le Roux, near Lubac. He came from a stay at a youth camp with his recent fiancée, a friend of his daughters. He was happy to find again the (relative) comfort of a house and normal food. Léonce took Hassler for some ascensions. When I came to Princeton after the Chapel Hill congress of 1957, Hassler and his young wife, Barney, lodged me for two days. They had a cute little girl, still a baby. Barney was nice and welcoming. I liked her and I thought Hassler was not an easy husband. Coming back home at the end of the afternoon, Hassler asked Barney if she had done what she had to do at an organisation, I don't remember which. Barney hesitated, Hassler got impatient, and Barney said, "I went but it was closed." Hassler looked surprised, but did not insist. When we were alone, Barney said, "I did not go, but he does not need to know everything, does he?" The Whitneys had a second daughter, but when I saw them again during a new stay in Princeton in 1973, they were virtually separated, and divorced soon afterwards. Hassler married for the third time at 83, three years before his death. I had met him a little before when leaving a meeting in the Academy, but he was already off as Lichnerowicz, who had seen him in Princeton, had told me. Whitney had been following De Rham without looking around. I advanced to say hello, De Rham told him, "It is Mme Choquet." Answer from Hassler: "Does she know I am divorced?", and he hurried to follow De Rham without another comment. Whitney was a very great mathematician. His book, "Geometric Integration Theory" remains a classic of rigor and clarity. It is on my bookshelves and I have consulted it several times with profit. In agreement with his passion for mountaineering, his ashes have been dispersed, following his wish, at the summit of a Swiss mountain — the "Dents Blanches".

Return to France

At the end of 1952, we received a letter from the University Education Director Albert Chatelet, offering Léonce and me a position in Marseille University. We were hesitant to interrupt our stay in Princeton, planned for two years, where we were quite happy. However, universities at the time had few professors and the simultaneous vacancy of two positions in the same department was rare. Moreover, a nomination in Marseille pleased us both. Léonce was born a southerner, and I thought of the happy memory that my mother kept of Marseille. Leray, thinking only of our interest, advised us to accept. Marston Morse promised us that if we wished it, the Institute would welcome us later to make up for our premature leave. We therefore accepted the offer of the French university and went back to France.

I went to say good-bye to Einstein. He greeted me as always very kindly, wishing me satisfaction in my future teaching job. This wish disappointed me because I had hoped for more joy from my future research than from teaching, an occupation necessary to a mathematician to earn his daily bread in a time where the CNRS offered permanent positions only exceptionally.

We benefited from another stay at the Institute of Princeton in the fall of 1955, but, then, Einstein was no more.

9

MARSEILLE 1953–1955

The University

After his thesis, Henri Cabannes was appointed Professor at Marseille University. Two positions in mathematics happened to be vacant there in January of 1953. Thanks to Henri Cabannes and Albert Châtelet who reigned then on nominations in universities, these positions were proposed to Léonce Fourès and his wife, that is, me. We had shortened our stay in Princeton to teach in Marseille.

The faculty of sciences had, then, about twenty professors. We doubled the number of mathematicians. The other mathematicians were Cabannes, of our age, a specialist of the mathematical aspects of fluid mechanics, and Vincensini, of the previous generation. Vincensini was a Corsican passionate about his discipline, geometry, and its teaching. Indeed the geometry familiar to him was somewhat outdated, very far from Bourbaki's, but it was from a glorious past: Darboux, Goursat, Elie Cartan, Levi-Civita, and we were not fans of Bourbaki. Vincensini was respectable as a mathematician, and as a man. Cabannes and us respected him; harmony reigned in the mathematics department. We all attended the seminars organized by any one of us, an opportunity to learn and to make friends outside of one's specialty. I particularly remember the visit to Marseille, organized by Vincensini, of the great Italian geometer Enrico Bompiani. Léonce and I invited him to dinner at our home. He was charming, an old man to us at the time. He had taken, on his knees, the young Michelle who asked him his name. His answer, "Enrico" surprised our daughter who said, "Haricot, c'est un drôle de nom." (translation: "Bean, it is a funny name.") It made the Professor laugh. Like many Italians of that time he spoke very good French, and said that Haricot would be his French name.

Léonce and I had, each, six hours a week of teaching students in the first year of university: Léonce in Math-géné (Mathématiques Générales), and me in the mathematics of MPC (Mathématiques, Physique, Chimie). A course of mathematics for physicists was created two years later. I was naturally given charge of it. A new colleague, Bodiou, replaced me in MPC. The auditoriums were of a reasonable size for about sixty students. With a few exceptions, they wished to learn and did not cause disciplinary problems to professors who had a minimum of authority, as was our case. I had the beginning of a problem only the year I had accepted to give a course in mathematics to first year medical students, a course which was not sanctioned by any exam. At the first lesson, the auditorium was packed. Ruckus began to prevail, noise, throwing of paper arrows… I left the lecture theater, telling the students to come to me if they wanted classes. They sent me a delegation asking me to resume my lessons, assuring me that the rowdies were more advanced medical students who would not come again. I resumed my course before an auditorium significantly reduced but quiet. This course, not followed by an exam, was later removed from the program.

I had been happy to find, in Marseille, my friends Henri and Madeleine Cabannes. Madeleine was always optimistic, persuaded of the goodness of God in spite of the problems caused by her second son who, she told me, was not "well come". The baby Jean Paul did not tolerate any dairy product, except Roquefort. The courage and perseverance of Madeleine ensured his survival. In fact Jean Paul was what was then called a mongoloid. In 1953, the chromosomal origin of this handicap, trisomy twenty-one, was not known. Madeleine spent considerable energy on her son, thinking that assiduous material and intellectual care would make him an ordinary man. She even took him for a Bogomoletz treatment, and persevered in her hope in spite of its failure.

Unfortunately, Jean Paul was a severe case. It is now known that the severity of the harmful effects due to the supernumerary chromosome 21 depends on the location where it is attached to the other two. To her deep regret, pushed by her husband, Madeleine had to resign herself to placing the adolescent Jean Paul in an appropriate facility in the interest of her other four children, all "well come". I will speak again later of Madeleine and Jean Paul.

We were also warmly welcomed at our arrival in Marseille by most professors (there were no teachers at other levels then) of specialties different from ours: mechanics, physics, chemistry, biology. These were happy times where there were no bulkheads between different disciplines. The dean, called Choux, pronounced Chouxe, was a man who did not lack authority towards students, but was amiable with colleagues and ready to support their legitimate wishes for schedules or useful supplies. We had pleasant and interesting relations with a number of our colleagues and sometimes their wives. We found, among them, true friends, in particular, Jacques Valensi and his wife Marthe. Jacques, just about fifty, reigned in the mechanics Institute and its wind tunnel. He was an imaginative scientist and a cultured man. Born in the great Safaride bourgeoisie, he had, however, attended a Jesuit secondary school, then he had entered École Centrale, the incubator of industry bosses. However, he had a researcher's vocation. After having done a thesis, he had become a university professor. His family of bankers had felt it as a comedown. It is true that the salary of a professor, even at the highest level, is a small thing in comparison with those of the professions practiced by his family. Jacques did not regret this relative poverty, but not being more appreciated by his father gave him complexes. Jacques was happy when he was elected corresponding member of the Académie des Sciences, but disappointed not to be elected full member as he deserved. But the number of members was very limited, and in life desiring something too much is often a reason not to get it. However, Jacques Valensi was an imaginative, hard-working scientist and also an excellent organizer, who has created in Marseille a fruitful Institute of Mechanics at the forefront of research. Jacques was also a sensitive man and a faithful friend. However, perhaps his complexes had led him to make some enemies. I remember a meeting of professors where Jacques complained about having had in his laboratory the visit of a colleague, Desnuelles, charged with making a report on his activity. Desnuelles said with good humor, "I don't see why M. Valensi complains, I wrote that his research was exceptionally interesting and his management excellent." Jacques' wife, Marthe, was a very intelligent woman, full of common sense. She was a doctor, Director of Hygiene in Marseille. She told me that Jacques would have liked to be an Academician to be able to go to his father's grave and tell him of his success. The Valensis welcomed us often at their home and became real friends to us. When they retired, they had a very pleasant house built in les Lecques and went to live there. I continued to visit them until their deaths. We also became friends with the Desnuelles couple, he, a very good chemist whom I met again later in the Académie, and she, a lawyer who became a magistrate when her children, growing up, left her enough leisure. I also got pleasure from my talks with a very good physicist, an optics specialist, Rouard. He was a cultured and amiable man who had known my father, and kindly congratulated me after correcting a paper of a student who had used a method I had taught to solve his problem. However, he added, with a smile, in the considered case it was akin to using a hammer to swat a fly. Rouard became Dean of the university when Choux retired. Both Desnuelles and Rouard, members of the Académie des Sciences, gave me their support when I was a candidate to this illustrious institution in 1979.

I also had the good luck to see, soon after us, a biologist colleague who was the only woman besides me in the Faculty of Sciences, Mlle Gontcharov, arrive in Marseille. She was quite likable and felt a bit lonely, suffering to see her works misunderstood. She experimented then on the transmission of information between cells. The memory, deformed and fragmentary, which I keep of it, is of many glass cups containing worms that she cut in two. She exposed some halves to light, then put them in contact with other non-exposed halves. It was very meticulous and repetitive work, for which I would have been very little gifted. I admired the courage and patience with which she was doing these experiments, compiling the results and persevering in the positive conclusions she drew, in spite of the skepticism of her biologist colleagues. The truth of an experimental discovery may be more disputable than that of a mathematical theorem, but I felt ready to be confident in the results of the very serious work of Mlle Gontcharov, an indisputably competent and perfectly honest woman. We became friends, supporting each other in the hazards of our lives. We both left Marseille the same year and, to my regret, we stopped meeting. I think she had bad memories from her stay in Marseille which she preferred to forget. However, she had the kindness to come briefly to Marseille in 1960 to testify at my divorce trial: divorce by mutual consent did not exist at the time. I deeply regret her no longer being here to receive the apologies that biologists should address her now for not recognizing the value of her remarkable work.

Private Life

Léonce and I, when leaving for Marseille upon our return from the US, had left Michelle with her grandparents in Albi, just enough time to find housing. Fortunately for me, but to the regret of my mother-in-law, it was not long. We found a nice furnished house at "la Pointe Rouge" near the beach, which we could rent for a year. We had the good luck to take with us a Spanish woman, a little more than forty years old. She was a widow named Pilar, whose only son was doing his military service. My parents-in-law had hired Pilar to help them take care of Michelle, who was then a little more than two years old. Pilar was perfect. We were very sorry when a year later she left us, in spite of her attachment to Michelle and, I think, her sympathy for me. She believed it necessary to be with her son as he had returned from military service.

Léonce, Michelle and I had a pleasant year in Marseille in 1953, living in Pointe Rouge with Pilar who relieved me of any house concern and sincerely loved Michelle, a cute and smart little girl. Her father accepted her existence and its consequences. He did not draw pleasure from it, but he did not complain.

At the beginning of the 1953 summer, before going as each year to Benagues, we had rented for Léonce, Michelle and I, and also Pilar, an apartment in Le Roux, a village on the national road 1.5 km from Lubac, where there were well-stocked bakery and grocery shops. Gustave had rented a house in the neighborhood — I don't remember exactly where. He was there with his three children and his sister Marthe, come to replace her sister-in-law who had left him. I did not know the motive at the time. I learned later, from Robert Bourgeon, that she had wished to replace her spouse, but that the substitute she had in mind, tempted one moment, in the end did not want to leave the wife he already had. I was very touched by the loving care of Gustave for his children. He was very nice also with mine, more so than her own father.

During a stay of Hassler Whitney with us in Le Roux, Léonce and Gustave made a few climbs with him. I sometimes went with them, for a part. The last one brought Léonce and Hassler in Italy across the Alps. I had gone with them up to a mountain hut. The following day, I started going down alone on the glacier as it was planned. I must confess that after a while, I began to be scared of falling into a crevasse. I hailed a lone alpinist, who descended blithely, and told him of my fear. He answered cheerfully, "OK, take the train!" and I followed him, reassured, to the terrestrial path.

At the end of the holidays, we moved to the apartment we had rented in the building called "le Corbusier". Our house at "la Pointe Rouge" no longer being available, we had had to look for other housing. The building of the famous architect, Le Corbusier, was then at the end of construction and its apartments leased, for a moderate price. Léonce happened to know the man responsible for the attributions of these lodgings, and we became the first occupants of one of the apartments of what was then called "the radiant city". It is now a historical monument. Léonce later bought the apartment, again for a moderate price and lived there until his death on 5 March 2015. Our apartment was on the seventh and last floor of this rather monstrous building. In fact, I did not much appreciate it in spite of the splendid view of the Mediterranean Sea offered at its balcony, but from our place, the sea was only accessible to birds. The building is on stilts, the ground floor under them is a garage for the cars, the remainder was a wasteland devoid of any amenity or even greenery, at least at the time. The roof terrace was supposed to serve as a pleasure garden, but it was bare, windy and, fortunately indeed, surrounded by a wall high enough to prevent accidental falls. That terrace had nothing tempting. Maybe it is different sixty years later. The apartments were located on either side of a corridor, and children living in neighboring apartments played in the electric light of this windowless corridor. However, Michelle keeps a very good memory from these games, especially the ones with her pedal scooter. A possible advantage of this le Corbusier building is that one could live in it without hardly ever going out, because one finds, there, shops selling vital necessities. For my part, I appreciated that it housed a preschool. Michelle liked the first year, now called, in France, "cours préparatoire". Her teacher, Mme Ogier, was a very kind young woman with whom Léonce and I had friendly relations. Michelle, who does not have an especially good temper, liked school less the following year when the teacher was more severe. She told me one day that she did not want to go to school again. When I asked for an explanation she said, "Mme B. (the teacher) told me that my Bs were not well made." Things still contrived amicably.

The university, located near the train station, is far from the Corbusier building, but it is possible to reach it by using a tram. I did not take it often, because Léonce and I had the same schedule and in general we went in his car. I say "his" car, then a Simca, because Léonce was very attached to it and would not have let me drive it for anything in the world. When, wishing some independence, I expressed the desire to get a driving license, Léonce said I was unable to learn how to drive. When I persevered in my will,. he said that if I got the license we would buy another car since there was no question that I drive his. This purchase did not take place during our life together.

The Ascent of Mont Blanc

Léonce, Michelle, and I spent part of the summer of 1954 in a rented apartment in Le Roux (Valgaudemard) like the previous year. Léonce and Gustave wished to make some climbs in the preferred region of true mountaineers, the Alps near Chamonix. The desire also took me to go see the Mont Blanc, and even to make its ascent. It did not require, I was told, special qualities of mountaineering, only to be a good walker and have a good guide. Léonce was one and he accepted to guide us, Gustave and I, to the summit of this highest mountain in Europe. We went to sleep at the shelter of the "aiguille du goûter", a long climb since the cable car did not exist yet, but I did not find it too hard because I carried little weight. In the early morning, our roped party, Léonce in front, Gustave at the back and me in the middle, left towards the summit. Some other parties did too, perhaps two or three, at good distances from each other. Nothing to do with the current influx. Léonce cut steps in ice when necessary, and the climb to the top was not too painful. The view from up there is splendid, but the icy wind which prevails there destroys its charm a little. I take pride now in telling people that I climbed to the top of Mont Blanc, although, by the usual way, it is not really a remarkable feat.

Travels

During my years in Marseille, I had occasion to make several interesting travels related to my profession, approved by our kind dean, M. Choux (pronounced Chouxe). The first was with Léonce: we both gave a talk at the "Institut Universitaire" of Tunis where J. M. Souriau, a student at the ENS one year before us, had a position as professor. I remember a pleasant carriage ride with Souriau, his charming wife and their two young children. Souriau, a fertile and original mathematician, was appointed Professor in Marseille, soon after I left. His wife had learned Arabic in Tunis — she became a researcher at the CNRS after a thesis on journalism in Arab countries.

Another trip I made while I was in Marseille was to teach differential calculus in Morocco. The newly created Rabat University was indeed an annex of that of Marseille, and professors thereof took turns to teach there. It is how, in two or three weeks, I taught half the course on differential calculus. I took advantage of weekends and local coaches to visit Fès and Marrakech, two very different cities, but both magnificent.

My most beautiful travel was in the summer of 1954, to Madagascar and La Réunion. I went there to chair juries of baccalaureates in Antananarivo and in Saint Denis of which the universities were, at that time, both attached to the University of Marseille. H. Cabannes had assumed these functions with pleasure the previous year. I also keep good memories of my trip in these faraway countries which offer beautiful sceneries. In Antananarivo, I was very well-received by the administrator and French fellow teachers, but I was shocked by their colonialist spirit. I particularly remember a meeting where only one member was of Madagascan origin, with a name typical of that country. I was scandalized to hear his French colleagues make fun of him by deliberately mispronouncing his name. I was ready to protest, but the Madagascan, with a blasé smile, beckoned me with his hand to do nothing. I understood the autochthons' revolt some years later, with much regret however, for all the suffering the fighting has caused. La Réunion, I think, was uninhabited when white men arrived there — the problems were not the same as in Madagascar. At the same time as me, a literary professor at the University of Aix en Provence, M., had come to preside over the exams juries. He was a specialist of ancient Greek, aged about fifty. He told me he had accepted these faraway presidencies to be able, upon returning, to stop in Athens and visit that city at last, the cradle of a civilization to which he had devoted so much study. He was already in Greece in spirit, studying an Athens map during his leisure. He accompanied me only reluctantly, and not always, to the excursions organized for us by the very kind director using the school's car. During our meals, taken together in the hotel or during our common trips, the Professor of Greek expounded on his thesis on Homer to me. It was quite interesting and added some poetry to the landscapes under our eyes. I remember a small lake where rays of sunshine were dancing, and M. quoting verses of Homer on "the innumerable smile of the waves". I met in our hotel in Saint Denis a Professor of Law, also on official mission. He had local acquaintances, in particular a planter who owned a small plane to fly over his vast properties. This autochthon proposed to my jurist colleague to take him flying over the volcano; the jurist offered that we benefit from the opportunity. M. declined the invitation, but I accepted with pleasure. One of the following mornings, my colleague and myself went to the given appointment. Seeing me, the planter seemed a bit embarrassed, he said he was not sure that his small plane could fly over the crater with the weight of three people, but he added amiably, "We will try." I think the jurist regretted his invitation: he shivered during the flight, perhaps for fear, but also because he was in a short sleeved shirt, not having foreseen that in a small drafty plane at altitude it would be very cold. Thanks to my mountaineering experience, I had dressed warmly and I was confident in the pilot, sure that he knew the volcano and his plane. The aerial ride above a molten crater is a beautiful memory. A less pleasant memory is from a visit to a leper colony lost in the mountains with a doctor colleague. M. had declined the invitation. The mountainous landscape was wild and beautiful, but the meeting with a leper, a schizophrenic moreover, was rather depressing.

Congresses

The first congress to which I participated after our arrival in Marseille was in Strasbourg. Lichné organized, in 1953, in collaboration with the professor in Strasbourg, Koszul, a colloquium on differential geometry. I did not give a talk, but I had the good luck to meet and sympathize with a great geometer of Chinese origin, a professor in California, S. S. Chern. I remember, with pleasure, our animated conversations as we visited Strasbourg together. Chern, about ten years older than me, was a warm and friendly man, though already a famous mathematician. I have always been happy to meet him later. Although perfectly acclimated in the USA, he chose to use the important prizes his mathematical discoveries had earned to support the scientific development of his country of origin. He subsidized the great Nankai University in Tien-Tsin. Among other things, he built, on the campus, a comfortable house which he bequeathed to the university. I had the honor to stay with Chern a few days when I went to give a talk in Nankai in 1997. It was a pleasure to converse with him and to evoke memories of the past. I was happy to see the solicitude with which his young colleagues treated him. I did, very willingly, a little report about Chern at the colloquium organized in his honor at the IHES, after his death.

In July 1955, I participated in the congress where relativists celebrated the fiftieth anniversary of Special Relativity, in Bern where Einstein's mind had given birth to this theory. Controversial at first, it was then universally adopted as a physical theory, not only as the elegant mathematics elaborated by Poincaré. Its discoverer unfortunately had died a little before, aged seventy-six, from an abdominal aneurysm on which he had refused to have surgery. From what I know, Einstein had said it was "disgraceful" for an old person to cling to life, words to meditate upon today. In spite of Einstein's absence, the planned congress took place, organized by a knowledgeable and friendly professor at Bern University, André Mercier. I was invited to give a talk, thanks to Lichnerowicz, I suppose. The famous physicist Wolfgang Pauli, Professor in Zurich, attended this colloquium. He was a fifty-year-old man, the father of Quantum Mechanics, discoverer of the Pauli Exclusion Principle among other things, and a collaborator of Einstein's in some fundamental works. He was a genius, a powerful mind, logical and able to overcome the prejudices which hinder scientific discoveries. He was also a man who did not embarrass himself with conventions of politeness and did not hesitate, like Molière's Misanthrope, to tell people what he thought of their work. At this Bern congress participated a respectable physicist, Achille Papapetrou, of Greek origin, then Professor in Berlin, and later, Research Director at the CNRS in France. I don't remember the subject of his talk, but I remember that Pauli called him "Papagallo", an unkind comment! Papapetrou was, however, a worthy scientist. His works are not spectacular, but serious and correct. In collaboration with another relativist, Majundar, he has constructed an exact solution of the Einstein equations, which proved important in string theory. Papapetrou was rather tall and thin, a bit harsh in aspect, but in fact a very nice man, and perfectly honest. I learned to appreciate him when we were both permanently in Paris and received foreign visitors.

Pauli did not attend my presentation, but he told me the following day that, given the report that had been made to him, he regretted not having come. André Mercier, the colloquium organizer, gave a reception at his home one evening for a few participants. I was invited, as well as Pauli. I thus had the opportunity to have, with Pauli, a fairly long conversation where he was very amiable. He told me that, the following August, he would come for a while at the Les Houches summer school and give a conference, which I would be welcome to attend since I was spending the summer in Chamonix. I wrote to Cécile DeWitt, the very efficient founder and Director of the school, asking her authorization for me to come listen at this conference. Cécile answered that only full-time members of a session could attend lectures given in a session of the school. When I told her that Pauli himself had invited me, she wrote to me, accepting my arrival for this conference. Finally, I had to leave the Chamonix valley before the passage of Pauli in les Houches, and I had no opportunity to see him again before his death three years later, at only fifty-eight years old, from pancreas cancer. It was a great loss for science.

New American Interlude

At the end of summer 1954, Léonce and I went to the International Congress of Mathematicians, in Amsterdam. Many mathematicians I knew were there: Irving Segal and his recent spouse, and his still-bachelor friend Georges Mackey, whom I befriended, now a Professor in Harvard. There was also Hassler Whitney, permanent Professor at the IAS who had been our friend since our stay in Princeton and his stay with us in the Alps. I told Hassler that we wanted to take advantage of the stay that remained available to us in Princeton. Hassler had an invitation sent to us, for the fall of 1955. Léonce was not enthusiastic, but he accepted, under the condition that we leave Michelle to her grandparents and that before going to work at the Institute, we cross the United States by car from east to west and back. I let myself be tempted because I always loved seeing new places.

At the end of summer we left Michelle with her grandparents together with the family help we had at the time, and took a boat to New York. Léonce took care of the purchase of a car, which we sold back to a colleague when leaving Princeton, and we started for the west coast. The crossing of the vast plains of central US was a bit monotonous, but an interesting experience for someone who knew only France and Italy. It appears that these Great Plains have been wrecked by intensive cultivation. What a pity.

Of course I was not authorized to drive. I filled the hours, sometimes reading, sometimes sleeping on the back seat. We ate in fast food restaurants and slept in motels, which were numerous and cheap at that time. What a wonder when we arrived to the Rocky Mountains and to the splendid national parks: the Colorado Grand Canyon, the Bryce canyon and other spectacular landscapes. The lonely crossing of the Death Valley is also an unforgettable memory. We then went up north along the coast, visiting the Yosemite National park and arriving at beautiful San Francisco. From there we went back east, without going further north to reach and visit Yellowstone. I knew its wonders only much later. To say the truth, I have forgotten details of our 1955 trip. I remember splendid landscapes, but I don't know if we saw them while going there or returning.

We arrived in Princeton for the beginning of scientific activities. Our house was, again, a small pavilion on the campus. Unfortunately, the friends from our first stay were no longer there. Irving Segal, now married, was back in Chicago, and the other members of our small group had found positions in different places. Fortunately Jean Leray was present again, although without his family. Mme Leray preferred henceforth to stay in France with her children who were pursuing their studies. She herself had resumed her position as Professor in the lycée Marie Curie in Sceaux, happy not to be reduced to housekeeper anymore and to have, she told me, personal happenings to tell her spouse. The always-intense work of Leray prevented him from boredom in Princeton, but he felt a bit lonely and came fairly often to have dinner with us, which pleased Léonce as well as me. An amusing memory is an evening when Leray, sitting on our couch, had just finished his dinner, fortunately a simple one, when appeared a young man looking for him because he had been expected at dinner in Morse's house. Leray, a very polite man, got up at once, excusing himself for being late, and went to eat a second dinner. I was told later that M. and Mme Morse had not known that he had already had dinner (or pretended to be ignorant of it) and had encouraged him to help himself generously. Leray had stalled only on dessert. We did not evoke with him these painful moments.

Other French scientists were staying at the Institute, in particular Gustave Choquet and his collaborator and contemporary, Jacques Deny, with whom he pursued important works on potential theory. Gustave had begun to be interested in this theory after discussions with his colleague in Grenoble Marcel Brelot.

This theory contains a lot of profound and subtle mathematics; its results apply to various physical situations. Contrarily to mathematicians of the Bourbaki school, Gustave, though a pure mathematician, had no prejudice against the use of mathematics as a tool for physics. The mutual attraction between Gustave and I increased during our stay in Princeton, cooling the friendship between Léonce and Gustave.

In 1955, Leray did not give regular classes. He was working on a new subject which would lead him to a whole series of important publications. I listened, of course with great interest to the few conferences he gave, but I did not find in them, at that time, matter for personal work. Following a conversation at a meeting in Amsterdam with the great Swiss mathematician Georges De Rham, I had seen that the method I had used to solve second order partial differential equations of hyperbolic type could be used to study solutions of more general equations, called ultra-hyperbolic. The quadratic form defined by the principal coefficients is, in this case, of any type. These equations don't have, as far as I know, a general application in physics, but their solution was a challenge which had awoken my curiosity. I attacked it during this second stay in Princeton. I found the right problem, besides those classically met, which could be solved in their case. I gave a talk on this subject at a colloquium on partial differential equations at Nancy University in 1956, and at the University of Maryland in 1957. Since then, mathematicians have studied much more general equations and formulated, for them, relevant problems. My early results are only a particular case, but precise and solved, of these general works.

We left Princeton at the end of December 1955 and returned to Marseille after taking Michelle back. I had missed her a lot. Michelle tells me now that she keeps a very bad memory of this stay with her grandparents. My mother-in-law, however, wrote to us regularly, sending photos of a pretty little girl with round cheeks who looked happy. Her school teacher described her as "intelligent and studious", the first qualifying adjective was given to her all along her schooling, but the second one, seldom before she entered university. Our brains make our memories through the experiences we live, but how they are sorted in the present and especially in the future is seldom the same for everybody. To console myself of the absence of my daughter, I bought her a number of pretty American clothes which I exposed on the couch of our living-room to the amusement of our visitors. We returned to France by boat, the last sea voyage I made to the US. The plane has become cheaper — now it is the transatlantic boat which is a luxury.

Gustave, a Fulbright fellow at Princeton University in 1939, had come to New York by boat. At the time he had been a young bachelor. His devoted father had dipped into his savings to buy him a tuxedo — he had been told it was necessary to participate in transatlantic festivities. In fact Gustave put the tuxedo on only once, but he had kept a wonderful memory of his crossing. Some fifty years later he could not, despite his little taste for lavish spending, resist the temptation to treat himself to a crossing of the Atlantic on one of these boats which still made it, for a golden price. Upon returning, he hardly gave me any comments, but it was clear that he was very disappointed. The boats, but also the passenger, were not the same as they had been in 1939.

Life in Corbusier Again

Having returned to France in the beginning of 1956, we went back, with Michelle, to Le Corbusier. It was winter and very cold in Marseille, especially in our apartment on the seventh floor which was heated with forced air. I called the maintenance service. Two men came, a manager and a worker. The worker climbed a ladder to reach the hole from which the forced air was coming and measured its force; he said, "It blows well". The manager answered, "But there is mistral" (mistral is a cold wind coming to Marseille from the Alps). Heated by the mistral, tired from the trip and the tenseness of the last weeks, I caught a serious bronchitis which lagged for weeks. Our friend, Jacques Valensi, advised me to consult a specialized doctor, a friend of his. I followed his advice. The doctor had me X-rayed and found fairly severe traces of the tuberculosis infection suffered years before. He gave me a treatment for the bronchitis but advised me to take a leave and spend a few weeks in the mountains. I followed his advice, accepted his prescription for a leave and reserved a stay for two, Michelle and I, in a small, not luxurious but comfortable hotel in Praz sur Arly, a mountain village about ten kilometers from the famous station Megève. Conversations with another boarder, a woman a little older than me who had come also to rest, and a few passing guests, brought me some distractions. Sunny days were many in number and I often walked with Michelle. Gradually I found my strength again. The innkeeper suggested that I hire a charming young girl of his friends for a few hours a day to walk and distract Michelle. I accepted willingly. Michelle also was pleased with this diversion. As for me, I returned during these few daily hours to my favorite distraction, mathematics. I wanted to do easy work. I thought of the so-called 3+1 decomposition of the space-time metric where only geometric elements linked with space sections and time lines appear. I had computed this decomposition in arbitrary coordinates a few years before, following a suggestion by Lichnerowicz in view of my thesis. These computations had only been published in a paper for the Comptes Rendus of the

French Académie des Sciences, and mimeographed as a talk in a seminar. My thesis, on the advice of Leray, had been on a more fundamental problem. Resuming in detail these calculus problems was a pleasant occupation for my free hours. I sent the redaction to the "Journal of Rational Mechanics and Analysis", recently created. Lichnerowicz was a member of the editorial board, and my article appeared already in 1956. The 3+1 decomposition was done again, under another aspect, by three American mathematicians, Arnowitt, Deser and Misner. It found many applications in the study of solutions of the Einstein equations. It is often called ADM decomposition, though most of the time it is used under the form I had given, with simplified notations defined by another American, James W. York, then a student of J. A. Wheeler. I have often collaborated later with this physicist, my junior by sixteen years, who was called Jimmy by his friends.

For the Easter holidays of 1956, Léonce, as usual, brought Michelle to her grandparents. We went, with the Defaixes, to ski in Megève. What I remember of this stay is periods of idleness where I basked alone in the sun, completing the return of my health.

Congress in Nancy

In the spring of 1956, I was invited to an international colloquium in Nancy on partial differential equations. I gave a talk on my results on ultra-hyperbolic equations obtained the previous year. A young participant, Pierre Cartier, whose office at the IHES I now share, made to me interesting remarks. Less pleasant was my encounter with an American mathematician, Lipman Bers. Then aged about forty, he was a professor at the famous Courant Institute in New York and very full of himself. I was with him and Fritz John, another professor at the Courant

Institute who was less pretentious. A French mathematician passed us and told me that the university's consulting committee had promoted me to third grade professor. I manifested my pleasure. Lipman Bers said wryly, "Let us congratulate Yvonne Bruhat for her recognition as a third-rate mathematician." John looked very embarrassed, thereafter he was particularly nice with me. I said nothing, but when a moment later, with an air of doing me a favor, Bers asked me to show him around the city, I simply said "no". I regretted not to have had the presence of mind to say to Bers, "I am touched by these congratulations coming from a second-rate mathematician".

I found, in Nancy, kindness from other colleagues, in particular Delsarte, a member of the Bourbaki group, with whom I made the train journey back to Paris. Having made the mistake of telling me that Mme Dubreuil, one of the few women having formerly entered the men's ENS, was being ambitious, he corrected himself by adding that the search for a beautiful theorem is a noble ambition.

First Congress on General Relativity

In January 1957, I was invited to what would be the first International Congress on General Relativity, the colloquium in Bern in 1955 now being called number zero. That congress was organized by Bryce DeWitt and his wife Cécile, in Chapel Hill, North Carolina, where Bryce was a professor. American law did not, at that time, allow husbands and wives to be professors at the same university. Though having all the diplomas to be a university professor, Cécile had to satisfy herself during several years with being the temporary replacement for professors who could not give their classes. This did not reduce her activity, which I always admired. Cécile is an outstanding organizer, the Chapel Hill Congress is an example of that. Cecile had convinced the US army to support our travels from Europe. At the time, the flights for the US were not direct. On the way to the US, Lichnerowicz, Mme Tonnelat — a former student of Louis de Broglie directing theoretical physics in Paris — and I made a stop in the Azores. I can still see us taking a light dinner behind a window that overlooked a beautiful sunny landscape. I regret that I did not visit those islands.

At the colloquium, the main specialists of the time in General Relativity participated. I was very well-received by this group of experts. Peter Bergman, the former assistant of Einstein and author of a book which remained, for a long time, a kind of bible for relativists, was an open, active and friendly man who always showed me much sympathy, despite our works having little relation. Peter was working, with little except formal success, on quantizing the gravitational field with methods used in electromagnetism. He has had many followers, including the great physicist Bryce DeWitt, but gravitation has eluded their most valiant efforts. The quantization of gravitation is still an open problem despite the appearance of clever new ideas. I was told that a colleague had reported that I do not believe in quantization. Like many reports of things said being made by someone else, this one is completely false. I think quantum phenomena are a physical reality, their importance, even at our scale, reveals itself more and more. Their mathematical formulation is a hard problem; it has been very successful in giving numerical results in remarkably good agreement with experiments for fields that obey linear or semi-linear equations, but the non-linear equations satisfied by gravitation resist. Also, there is, at present, no experiment where a particle with the name graviton manifests itself. Being a reasonable woman, I have not tried to quantize gravity. I don't regret it, in spite of the saying by, I don't remember who, "Progress arrives by the unreasonable one."

An important participant at the 1957 congress was John Archibald Wheeler, a well-known physics Professor at Princeton University, who was new to the subject but full of imagination and contagious enthusiasm. He gave the name "black holes" to strange objects whose existence was then hypothetical. The shock of two of them created gravitational waves, the first observation of which made the headlines. John Archibald showed interest in my results and invited Lichnerowicz and I, to stop a few days in Princeton before going back to France. We met frequently thereafter, in the States or in France. We became real friends, also with his wife Janet, an intelligent woman full of common sense who made, with him, an enviable couple. Younger participants, very friendly with me, have since become famous; for instance, Charles Misner and Stanley Deser. Charles, a student of Wheeler, created an important group in the University of Maryland and collaborated with John Archibald and another relativist, Kip Thorne, in the writing of a voluminous treatise on General Relativity called MTW, nicknamed "telephone book". It has been on the bookshelves of every relativist. I particularly appreciate Stanley's introduction with Dieter Brill of the notion of the mass of a gravitational field as a global property of space-times which become like an Euclidean space at space infinity. He became Professor in Boston at Brandeis University.

It is in Chapel Hill that the idea of creating the General Relativity and Gravitation society germinated. It took shape and statutes in 1971.

To subsidize our stays, Cecile and Bryce had found an industrialist whose fortune had come from patents of his invention for refrigerators, Agnew H. Bahnson. This billionaire was passionate about science. He attended lectures and noticed, he said, my attentive and interested air motivating his interest in me. I have no prejudice against non-scientists interested in science; my mother was one of them. Unlike Bryce, with whom Bahnson's comments got on his nerves, I discussed willingly with this enthusiast of scientific discoveries. At the end of the congress, he invited the three French people, Lichnerowicz, Mme Tonnelat and me, to go with him in his plane, which he himself piloted, to spend the week-end in his home in a neighboring city. Mme Tonnelat, having little confidence in this transportation, declined the invitation. Always ready for new experiences, I wished to accept. Lichnerowicz accepted, in order to be my chaperone, he said. A chaperone had not been necessary because Agnew was married to a charming woman who received us very kindly in a vast and comfortable home. It was an interesting experience for me, this flight in a private jet and stay at a billionaire's. Agnew Bahnson was certainly a remarkable man; he also wrote science-fiction novels. Unfortunately, he drank too much whisky, and after a certain hour, he lost some of his lucidity. It was, perhaps, the cause of the accident which provoked his death in his plane a few years later. Bahnson bequeathed, to the University of Chapel Hill, a large sum to finance research in General Relativity. After the departure of the DeWitts for Austin, these funds were made available to the successor of Bryce, James W. York. It is through these funds that I could stay in Chapel Hill several times and continue, with Jimmy, an interesting collaboration. After leaving Chapel Hill and stopping a few days in Princeton, where I saw again my friend Hassler Whitney, Lichnerowicz and I went to Washington; our military plane for Paris left from there. Leray knew a professor at the University of Maryland who was a specialist of partial differential equations, Weinstein; he invited me to give a talk at his seminar. He invited Lichnerowicz and me to dinner at his home. His wife told us that she did not invite Marcel Riesz, also passing through Maryland, because he drank too much. That professor, born in Hungary where alcohol is appreciated, had apparently drunk a whole bottle of Armagnac left within his reach during a previous visit. Marcel Riesz, who had been a member of my thesis committee in 1951, had come in the afternoon to listen to my talk on ultra-hyperbolic equations. When Lichnerowicz and I came back from dinner to our hotel, around midnight, Riesz was sitting on a sofa in the lobby. He let Lichnerowicz go, but stopped me, saying he had been interested in my talk because in the study he was doing, on some equations presenting invariances, he had come across ultra-hyperbolic equations. He wanted to know if my results could be useful in his work. He started to explain to me his equations and the problems he encountered about their solution. I had had a busy day and it was past midnight — I just longed for my bed, but Marcel Riesz, some forty years my senior and having had, perhaps, a bottle of Cognac in addition, was as fresh as a roach. We had a fairly long, though inconclusive, discussion. Marcel Riesz would have gladly extended it but it seemed impossible to me to stay awake longer. I have always been a good sleeper; I am happy of it now.

Our military plane brought us back to Paris after a stop, this time in Newfoundland where there was a snowstorm that discouraged us from going out. Our Israeli companion, Nathan Rosen, Professor in Haïfa, who was more daring, tried an exit but was soon back. Transatlantic flights are now non-stop, it is a pity in a sense.

Holidays

Through our friendly colleague, the chemist Desnuelles, Léonce and I got to know a couple who lived in Fuveau, a small city on the road to Aix en Provence. The husband Louis Defaix, aged about fifty, was a doctor and a seasoned alpinist. His wife, Simone, about ten years younger, was without a profession, but the director of the household and their two pre-teen girls. Without being, herself, a real mountaineer, Simone loved mountain walking and skiing. She could not stand leisure one-on-one with her husband. She needed an activity shared with friends. She organized meetings and activities with energy and efficiency. Anxious to make our life easier and increase our leisure time, Simone found full-time maids for us. The first, Olga, was engaged to a military man and awaiting, patiently, his return. Simone ironically called him "Désiré" (desired) and doubted that the wedding would ever happen. It eventually did and when Olga left us to marry, Simone found a replacement, Odette, capable and serious but morose, the consequence, I think, of an unhappy childhood. I regretted the departure of Olga, who had been perfect for me and for Michelle.

The Defaixes and Léonce had become friends quickly. I did not mind it, because I shared their liking of walking, skiing and mountain expeditions, though I was not a real mountaineer. For Christmas or Easter holidays, Léonce brought Michelle to her grandparents and we went to ski with the Defaixes. At Christmas we went to Val d'Isère, Léonce and me in a hotel, the Defaixes in a rented house. For Easter the destinations were more varied. I remember a ski touring in the Austrian mountains across magnificent landscapes, with night stops in shelters perched high where the view was splendid. I remember also, a stay in Breuil-Cervinia, Italy, where the sun was shining but the ski run was icy and so steep that I could never arrive to the bottom standing up. During the summer holidays, I refused to be separated from my daughter. We rented a cottage in the Chamonix valley, the paradise of alpinists. Ascents made by Léonce with Louis Defaix or another of his partners were not within my reach. I sometimes accompanied them to the refuge, but it was a little distressing that what was, for me, a difficult and beautiful excursion was for Léonce only a necessary chore before the climb of the next day.

Every Sunday of the school year we went, without any possible alternative, rock-climbing in the calanques (creeks near Marseille) with the Defaixes, and sometimes other climbers. Simone left her daughters in Fuveau, telling me she saw them enough during the week, but I always took Michelle with me. The calanques of Marseille offer splendid landscapes, and one could freely circulate there at the time. However I never appreciated rock-climbing as a specialty in itself, though I gave it a try sometimes. I don't remember if and at which level Simone practiced it.

For summer 1956 we rented an apartment near Chamonix so that Léonce could satisfy, with Louis Defaix, his passion for mountaineering. My parents-in-law came to see us, to spare me the trip to Bénagues. For the end of the summer Léonce organized for us a journey in Greece: traveling in our car until Brindisi, where we would cross the sea on the deck of a boat with a stop in Corfu where passengers and fowl embarked, and an arrival in Patras. Then it would be followed by a trip within Greece in coaches with stops in Athens, Olympia, Mycenae .... Our only luggage was a rucksack. Of course it was a beautiful trip. The Olympia site, especially, at the time a green meadow sprinkled with ruins, leaves in my memory very beautiful images. It seems to me that Léonce was more occupied taking specialist photographs than he was sharing my emotion before these ruins of a prestigious past. This journey did not fill the gap between Léonce and me.

My Couple Deteriorates

We had met and married very young. Léonce in the beginning had been distraught by the separation from his family to which, as an only son, he was very close, and I had been upset by the disaster that had befallen mine. I am grateful to Léonce for the support he had been for me during these painful years. So we were attached to each other, then married, without being well-tuned. I cannot blame Léonce for the bad reception he had given to his daughter, nor for his refusal to have other children, since he had told me before marriage that he did not want any. However, his behavior and various features of his character, which revealed themselves with the years, had distanced me from him, as features of mine distanced him from me. Our married life was only a juxtaposition.

Mme Desnuelles, whom I met in the Academie several years later, at a ceremony in the honor of her prematurely deceased husband, told me he had felt remorse to have introduced us to the Defaixes, thinking that they had been responsible for my divorce with Léonce. It is partly true. I did not feel hostility against the Defaixes. I found Louis rather pleasant and I was not jealous of Simone, a not particularly attractive but sensible-seeming woman ten years older than me, who was attached to her husband at least by interest. However, their quasi-constant presence in our leisure time, the important place they took in Léonce's life, without my sharing the satisfactions he got from it, played a sure role in the degradation of our married life.

10

TRANSITIONS 1957–1964

The Congresses of the "Mathématiciens d'Expression Latine"

During the summer of 1957, Léonce was invited to the US in a program of his specialty, Rièmann surfaces. Not having been invited, I remained in France. During the time Michelle spent with her grandparents, I went to the first Congress, of mathematicians speaking a Latin language, in Nice, where reigned then the famous mathematician Dieudonné. The association of Latin-language-speaking mathematicians had been initiated by André Lichnerowicz, to whom the English language never became familiar. His wife Suza, a high-school Spanish teacher, spoke English perfectly. André told me he had made no effort to learn English to leave this superiority to Suza. He was a loving husband.

Gustave was at this congress in Nice, without his family. His instructor and collaborator on potential theory, Marcel Brelot, was also present with his wife, both very friendly, and their daughter Claude, a young girl who showed the independent character usual for her age. I remember that we went together, also with the Lichnerowiczs, to a striptease show that connoisseurs found devoid of artistic value. Gustave and I, good walkers, had a pleasant excursion in the vicinity of the city.

Another one of these congresses took place in 1969 in Bucharest. Romanian is indeed a Latin–derived language, the closest to French, according to some. Before the war, there were very good Romanian mathematicians; several had been trained in France and spoke perfect French. They were still teaching at the university, as relations between the French and Romanians were excellent. The congress they had organized, with the collaboration of Lichnerowicz and Dieudonné, was very successful. That was before the dictator Ceaucescu had organized the persecution of intellectuals, though he had a mathematician daughter who had tried to protect her colleagues. In 1969, Bucharest was still a city full of charm with old monuments, which was then unfortunately destroyed by the misguided ambitious renovation of the dictator. I spent an evening and a night in Bucharest in 1982, on the way to Shanghai. I was very sad to see the wreck suffered by the city. The Romanian police were now severe. They wanted to force my fellow travelers and me to spend the night on the airport benches. Fortunately, a young Romanian woman, charming and efficient (I don't remember what her official functions were, but she had the kindness of Romanians of the past) was able to take us to spend the night in a city hotel. I did not recognize the Bucharest of 1969. It is said that the new authorities do their best to remedy this disaster. The mathematics department regains life, though the best Romanian mathematicians have fled under the Ceaucescu dictatorship.

The last congress of Latin-language-speaking mathematicians, also very pleasant, was organized in Coimbra by Antonio Ribeiro Gomes, I will speak of him later. At that congress, I sympathized particularly with a Brazilian mathematician, J. Palis, who became an associate member of the French Académie des Sciences. Palis, like the other participants, spoke perfect French. Unfortunately, the congress in Coimbra was the last for the Latin-language-speaking mathematicians. Nowadays, mathematicians from Italy, Portugal or South-America do not learn French at school but English. Perhaps it will be Chinese for the next generation.

Separation, Stay in Les Lecques

After the congress in Nice, I got Michelle back and went with her first to Dammartin where my mother, my brother, my sister and her family spent part of the holidays. I then went to Les Lecques where I had reserved the first-floor apartment of the house of Miss Fabry, the aunt of Henri Cabannes who had rented out the apartment. A neighbor in Le Corbusier with whom I sympathized, herself divorcing, advised me to do the same and recommended I go to see the lawyer who had taken care of her case. I did. The lawyer said that the behavior of Léonce, in particular his refusal to have another child, was admissible grounds for divorce. Divorce by mutual consent did not exist then; you had to prove the wrongs of the spouse. I preferred to remain in Les Lecques rather than return to the apartment of Le Corbusier. Following the advice of the lawyer, I filed an application for separation. He asked for a partial payment which I gave, without foreseeing the first of many others, and that this recommended lawyer would enjoy dragging on this divorce trial. Léonce was shocked at first by this request for separation, a very human reaction: one does not like someone else taking even what you throw out. He then accepted the idea of divorce, refusing, however, to assume all the faults alone. I agreed. Obviously he did not ask for custody of his daughter. I did not care for the alimony; my salary was sufficient. However, while I had witnesses of the wrongs of my husband, he had difficulty in finding testimonies on mine. My lawyer was delighted to prolong the trial. In 1960, our friend Jacques Valensi came to my aid by sending me to an attorney of his friends. That one took the matter in hand, made me write to Léonce a letter of insults, and our shared faults divorce was quickly pronounced in the presence of my witnesses, the absence of a lawyer and Léonce. I had custody of Michelle but her father was entitled to take her for part of the holidays. The Defaixes, really devoted friends, bought a house in les Praz de Chamonix so that Léonce could receive his daughter without depriving himself of climbs and ski in their company. The victims of this arrangement were the parents of Léonce who ended up seeing less of their granddaughter. Michelle tells me now that she keeps bad memories of these holidays, at least in winter, because of the compulsory skiing in any weather, but she did not complain too much at the time.

For the school year 1957–1958, I kept teaching in Marseille and residing in Les Lecques. I had enrolled Michelle in the public school of Saint Cyr, two or three kilometers from Les Lecques, in ninth since she had been in school two years in Marseille. The headmistress remarked that my daughter, born in 1950, was young for that class. But, fearing that the upheaval of her life would hurt Michelle's studies, I insisted for her registration there and I said without thinking, "But she was born at the end of the year, in November." This stupid argument led to the admission of Michelle in third grade. Going from grade to grade, my daughter got the baccalaureate at the early age of 16. I blame myself for this sometimes, in spite of the protests of the one concerned.

Mlle Fabry's apartment was already rented for September, but the Cabannes had found for me a large and comfortable house called Dar el Chott next to the beach. It was the property of a cousin who only occupied it in summer and rented it out off-season for a very reasonable price. That house has now been transformed into a hotel called "Chanteplage". I settled with Michelle in Dar el Chott, and thus lived some 300 meters away from my friends. Madeleine had found, for me, a responsible and friendly lady, aged about fifty, who took over the housekeeping and the care of Michelle when I had to go to Marseille for teaching. I was away 24 hours per week, taking a train at the beginning of the afternoon on Tuesdays and coming back at the end of the morning on Wednesdays, after having given my two two-hour-long classes on mathematical methods of physics. I slept on Tuesday nights at the hotel, in peace to prepare my lessons. The Cabannes already had, besides Down syndrome sufferer Jean-Paul, two boys: Jean-Pierre — a few months older than Michelle — and André — a few months younger — and a girl of little more than a year, Hélène. In the fall of 1957, they had a fifth child, a charming baby, Benoit. It was the time of the baby boom. Madeleine, a devout catholic, would still want other children, but as she had to have surgery for a fibroma, it stopped her pregnancies. Henri was rather glad of that; he said, "Five is good." Madeleine was helped by a very nice girl living at their home, but her baby, the little Hélène and Jean-Paul did not let her idle. The two elders went to the public school in Saint Cyr, but still demanded a lot of time from their mother. Madeleine came each morning to take Michelle and drive her to school with her sons. Jean-Pierre and André were half-boarders but I picked Michelle up to have lunch together. I did not have a car but a bicycle, and the trips les Lecques — Saint Cyr with Michelle sitting on the luggage rack were, for me, salutary exercise leaving me good memories. In general, Madeleine went in the afternoon to pick up the children from school. Often, André remained for some time at Dar el Chott to play with Michelle, and we brought him back home on foot.

During my free hours, I devoted myself to my favorite distraction, research in mathematical physics. I worked in the spacious living room, on the ground floor, with a large window separated from the sea only by the beach. Michelle and I took our meals there on a table, just far enough from this bay window to be sheltered from the sun. It was wonderful. It is also at this table that I worked when Michelle was at school. Since then a concrete walkway, slightly raised, has been built, separating houses from the seashore.

Research: Relativistic Fluids

In my thesis I had studied the evolution of space-time with no non-gravitational energy sources other than electromagnetism. In reality, there are obviously many others: first, simply matter. In classical General Relativity, energy sources are represented by a symmetric 2-tensor, that is, once coordinates for space and time are chosen, ten functions of four variables. In the simplest representation of matter, called dust, the tensor source is determined at each point of space-time by the matter density and the velocity of an infinitesimal element that is a scalar function and a vector field. In a more realistic representation, matter is a fluid: to the previous unknowns, one adds the pressure. The problem of fluid motion has been studied for a long time in Classical Mechanics, though many problems remain open. In General Relativity the problem is complicated by the fact that the space-time itself is determined by its content. Equations governing fluid motion must be coupled with Einstein equations. In 1957, a number of important works had been dedicated to the study of the evolution of classical fluids, but there was not a general theorem on the evolution of relativistic fluids. To know if it obeys causality was an open problem. It was natural that, given my anterior work, I was interested in this question.

I explored the literature concerning the generic evolution of classical perfect fluids (that is, non-viscous). I found, in it, no general theorem concerning their evolution through time. It was, of course, already known that this evolution can only be regular for a finite time, because of shock formation. Viscosity prevents the apparition of shocks. Research on global existence — that is, for an infinite time — of a regular evolution of viscous fluids, governed by the so called Navier-Stokes equations, had given birth to remarkable work by my master Jean Leray. I knew that relativistic viscous fluids would lead to fundamental conceptual problems of propagation with infinite speed, incompatible with relativistic causality. I wrote to Leray, asking him for information about perfect classical fluids. He soon sent me a demonstration of his own of a local evolution theorem for such fluids, using the definition of vortices and the so-called Helmholtz equations they obey. Synge and Lichnerowicz had proposed a dynamical definition of vortices in curved space-times and proved equations analogous to the classical Helmholtz equations. Using this definition and inspired by the equations written by Leray for classical fluids, I succeeded in proving that the Einstein equations, with a perfect fluid as source, coupled with the equations satisfied by such relativistic fluids, are a hyperbolic system in the general sense I had learned from Leray's course. Propagation was causal under a simple condition, physically reasonable, on the relation within the fluid between density and pressure. Leray presented, to the Académie, the paper I sent him on the subject, refusing to be quoted. I think, however, that the theorem on classical fluids was not written anywhere at the time. The article developing my proof appeared in "Bulletin de la Société Mathématique de France"; Leray was then its editor. I found, later, a direct demonstration which does not use vortices, considering, also, the thermodynamic properties of relativistic perfect fluids introduced by Abraham Taub. However, I keep a very good memory of the work done in Les Lecques behind the window of Dar El Chott and of the handwritten letters exchanged with Leray in 1958.

Single Life in "Résidence de Tourvoie"

At the end of the school year, I had to return the use of the house to its owners. In fact I wished to leave Marseille and go back to Paris. I did not hope to be, at once, appointed to the Sorbonne, which was then a privileged university; its professors even had a salary higher than the provincial ones. However, a university institute had just been created in Reims and was run by the Paris university. Nominations to this institute were made by the Dean of the Faculty of Sciences of Paris, Joseph Pérès, the president of the jury of my thesis. I went to see Pérès and obtained the professorship in Reims. At the time, university professors had only three hours of class per week. I could give them in a day, taking an early morning train to go there and a late afternoon one to come back, having given two hour-and-a-half classes separated by a lunch break. During the first semester I taught General Mechanics, during the second, Continuous Mechanics, inspired by a recent book of Paul Germain. The premises of the institute were provisional, they included neither library nor offices for the professors, and therefore provided no motivation to stay after the classes. It was perfect for me. The great officer for mathematics at the time was a former ENS student, Belgodère, helped by a devoted secretary, Mlle Lardeux. Access to the mathematics library shelves, located in the Poincaré

Institute, was in principle reserved to professors in Paris. To benefit also of this privilege, I argued that the university institute where I taught depended from the Paris University. Belgodère, whom I knew very well, could hardly refuse, but he told me he was embarrassed to grant me this favor because he should then also grant it to the Tunis University Institute professors and he did not know them enough to trust them. Tunis is farther from Paris than Reims. I disregarded Belgodère's scruples and I went fairly often consulting mathematical documents in the Poincaré institute; the Internet had not existed yet.

With the financial help of my mother, I bought a small prebuilt pavilion in a recent building plot called "Résidence de Tourvoie" in Fresnes, a suburb south of Paris. There was only a sixty-square-meter ground floor composed of two rooms with a kitchen, bathroom and toilets. One of the rooms, fairly big, was separated by a curtain to become a bedroom for me and living-room for both of us. A local carpenter had made shelves for us that occupied a wall. The other room, long and rather narrow, served both for Michelle to sleep and for me to work, with a desk behind the window and a board fixed on the wall for scientific documents. In front of the pavilion, a lawn and a few flowers I had planted embellished the place. The residence is a fifteen-minute walk from the subway Antony, from which I joined the East Railway station and Reims. An amusing memory: on the train, one day, I entered in conversation with another traveler. I told him I lived in Fresnes and taught in Reims. He was beginning to pity me for such a difficult life, when I told him that I spent only one day a week in Reims. He said, "What you teach must then be extremely important." It was flattering!

I had the good luck to find, in Fresnes, like the previous year in Saint Cyr, a lady of about fifty years, friendly and completely reliable, to help me in housework and take care of Michelle in my absence. Mme Teilhet was the wife of a policeman. She told me her husband preferred to work at night, thus avoiding giving tickets, a task he did not like! I enrolled Michelle in the public school. She went there easily on foot, with a companion who lived in a nearby building. Her teacher was intelligent and friendly, and Michelle liked her. She had made friends in her class, and in the residence, with two sisters living just opposite us. The sisters, Marie-Françoise and Marianne, helped their parents care for babies for whom they were fostering.

Relativistic Fluids Again

As for me, I continued research in mathematical physics. Henri Cabannes, who worked on classical fluids mechanics, had pointed out to me an article by two physicists, Hoffman and Teller (one of the fathers of the atomic bomb), who studied fluids with infinite electric conductivity in Special Relativity, without however, considering the most general case nor extending their study to General Relativity. Fluids with infinite electric conductivity play an important role in plasma physics. Their mathematical representation leads to interesting original equations. After studying the Hoffman and Teller article, I looked for the most general equations and their coupling with Einstein's ones, in view of applications to astrophysics. Lichnerowicz resumed this study a few years later and completed it by using a result about partial differential equations, just obtained by Leray, which I had told him about on the telephone. Lichnerowicz gave a course on relativistic fluids at Dallas University (Texas) in 1967, which was published in the USA. He quotes me in the bibliography, but not in the text. I held a grudge against him for some time for that. The misfortune which struck him a little later, the suicide of his eldest son Marc who was 24, made me forget this small grievance. Marc Lichnerowicz had just completed an interesting PHD on the application of thermodynamics ideas to problems of the economy. He should have enjoyed a brilliant future, but a serious illness in infancy, with a very high fever, had caused him a nervous imbalance. He, and consequently his parents, suffered from it until his tragic death. The Lichnerowiczs managed to overcome their misfortune with the help of their two other children, Jacques and Jérôme, who became, respectively, an architect and a doctor.

The Illness of Michelle

I had a serious problem with Michelle's health after the Easter holidays of 1959. She had, before these holidays spent with her father, a tonsillitis, apparently cured when she left Fresnes. Upon her return she had what the local doctor, G., confused with appendicitis, sending her in emergency to the nearest clinic where she was operated on without delay by the surgeon attached to the establishment. Soon back at home Michelle complained of pains in her two wrists. I had been surprised by the appendicitis diagnosis but I had trusted Dr. G., who had been appreciated by my neighbors. However, I had some medical knowledge and the pain in the wrists made me think of a different condition, A.R. (Acute articular Rheumatism). Dr. G. confirmed my diagnosis, without invalidating his previous one. Lichnerowicz advised me to consult his friend, the pediatrician Lepintre, who had followed Michelle as a baby. Lepintre, who agreed to come from the sixth arrondissement of Paris to Fresnes, confirmed the diagnosis, and ordered penicillin and cortisone to prevent heart damage but said to only start a few days later because of the operation of this false appendicitis. He advised me to ask the local doctor to visit Michelle daily as he could not do it himself from Paris.

Two days later Dr. G, listening to my daughter's heart, observed an anomaly and advised to start, at once, the cortisone, ordering also, an appropriate diet without salt or fat. Dr. G. told me later he had almost let another child die of appendicitis, because before sending him to the hospital for surgery he had waited for confirmation of his diagnosis by an emergency call from the parents! The eldest daughter of my friend Nicole had suffered an A.R. with cardiac damage as a sequela, a lifelong handicap. I was very worried for Michelle and I took the excuse of illness sometimes when I should have gone to teach in Reims. The doctors, G. and Lepintre, came again several times. Michelle did not return to school that spring. She had to keep to her bed for a long time. To divert her attention, I asked Mme Teilhet to bring her a kitten. Michelle was delighted and called it Moustique, which became soon Moumousse. Finally this episode ended without sequela for her. Only my optimism bore marks.

In September 1959, Michelle was admitted, without difficulty, in seventh, the last year of primary school, in spite of her long absence the previous year. The entrance examination in sixth, which I had failed in my youth, did not exist anymore. However, there were still some requirements for college (secondary school) entry. Sixth, and the teachers of seventh were rather demanding of their pupils. Michelle bore it with difficulty. Whether due to stress or microbes, Michelle began to show signs of relapse. Her sedimentation, which I had regularly measured by doctor's order, took a value greater than normal and she had to undergo a penicillin treatment again. Lepintre did not this time order cortisone but daily aspirin, to be stopped when the sedimentation speed returned to normal. I decided, however, not to send Michelle back to school, but to enroll her, as she wished, in correspondence courses. I found a student, Mlle Faurie, who came regularly to help Michelle do the work. This girl was perfect — Michelle worked better with her than at school, and with pleasure, regularly sending homework to correspondence education. At the end of the year Michelle was admitted in sixth, not yet ten years of age. Mlle Faurie told me that perhaps, given her youth, it would be better that Michelle made another seventh. She would not have time to take care of it, but she could help me find a replacement, I could also enroll her again in the public school. Michelle refused emphatically to stay in seventh for another year. I hence enrolled her in the nearby middle school in Antony. Her year of sixth was without problem.

Michelle had no A.R. relapse. However, I did not stop worrying about her health. She followed a preventative prescription of low doses of penicillin until she was fourteen. She is now allergic to this family of antibiotics and reproaches me for this treatment. I cannot blame myself, for I followed the advice of Dr. Lepintre, a pediatrician. And moreover, without telling Lepintre, I took Michelle to see a professor in the faculty specialist of A.R. I can do no less since I met a woman suffering from debilitating cardiac complications, the after effects of a badly–attended–to A.R.

Gravitation, an Introduction to Current Research

In 1960 (or was it 1961?) Louis Witten, the father of the famous physicist Edouard Witten, himself a valuable relativist and a pleasant man whom I had met at the Chapel Hill congress, suggested I write a chapter on the Cauchy problem in General Relativity for the book he intended to edit, entitled "Gravitation, an Introduction to Current Research". I accepted with pleasure. It was my first scientific text in English: at that time French was still a language accepted by all scientific journals; it had even partially occupied the privileged place which English has now, probably temporarily, as everything in the universe is. I had established friendly relationships with the young and brilliant Stanley Deser, whom I met in Chapel Hill in 1957. Stanley spoke perfect French as he had done part of his studies in Paris. He offered to translate, in Anglo-American, the chapter which I would write in French. He advised me, apparently selflessly, to write in detail my work on elementary solutions and propagation of gravitation and not to include the generalization I had made of the Darmois-Lichnerowicz 3+1 decomposition. His warm kindness convinced me to follow his advice — my chapter "The Cauchy problem", published under the name Yvonne Bruhat, does not refer to this decomposition. I included in it an original way to solve the constraints as they appear for the Einstein equations in harmonic coordinates. This method has been little used; perhaps it is a pity. The book edited by Louis Witten, rich in topical content, saw great success. I should perhaps be grateful to Stanley who has translated my chapter, but this gratitude is dimmed by the writing for the same book of the chapter "The Dynamics of General Relativity" by Arnowitt, Deser and Misner which, in its beginning, resumes the 3+1 decomposition without even quoting my work on the subject. This chapter of the book, edited also by Louis Witten is very often quoted, for its beginning in fact. My chapter is seldom quoted. I don't complain about this because my original works on its subject are quoted each time they deserve it.

The Royaumont Congresses

When I lived in Fresnes, my mother, recently retired, used to come for lunch with Michelle and me on Thursdays, which were school holidays. We had a good time because my mother, when she was not unhappy, was pleasant company, interesting and understanding. In the spring of 1959, the third Congress on

General Relativity took place in the ancient abbey of Royaumont, which became an international meetings center. It was called GR 2, jointly organized by A. Lichnerowicz and M. A. Tonnelat. My mother accepted, without problem, to come help Mme Teilhet take care of Michelle while I went to this congress. I think it is the only time she took charge of one of her grandchildren, but she did very well, though it is true that she was helped by Mme Teilhet.

The conference brought together nearly a hundred and fifty participants, an important number at the time for a specialized colloquium. I see, on my list of publications, that I talked in Royaumont of my work on charged fluids. The only memory I keep of my lecture is the vexation I felt when I saw Fock, the head of the Russian relativists sitting in the first row, take off his hearing aid in the middle of my talk. Fock was an overbearing man, imposing also in his size and weight. I still have on my bookshelves his book on General Relativity; it is certainly much work, but it did not teach me anything. Fock's belief in the physical meaning of particular coordinates, called harmonic, disagrees completely with the General Relativity conceived by Einstein. A physicist who was, for some time, an authority in China, Zhou Peiyuan, shared Fock's belief, but not his intransigence. Zhou Peiyuan was otherwise a very intelligent and understanding man who had known how to keep the esteem of his fellow scientists through the Cultural Revolution as Zhou Enlai in politics. Coming back to the Royaumont congress of 1959, I see, on the list of participants, the names of many scientists from various countries who have contributed in important ways to our understanding of General Relativity and to its applications in physics and astronomy. I have personally known most of them. Some were young in 1959; others less so. Some names have fallen into oblivion; others have acquired an international reputation or already had it in 1959. I had met most of those at the congress of Bern or Chapel Hill. A scientist whom I had not met before, who had come to Royaumont is the famous Paul Dirac. I had come from Fresnes with my "deux chevaux". Lichnerowicz gave me the task to fetch Dirac at the train station. This famous scientist sat next to me but didn't tell me a word, though he knew I was a scientist, not a professional driver. His attitude did not change during the meals that we had occasion to share. On the other hand, his wife, who accompanied him, was a charming woman who was trying to be kind for two. The conference given by Dirac left me no memory. I see on the colloquium proceedings that it was on the energy of the Einsteinian gravitational field, without bringing much on this difficult subject.

Quite a few of the participants to the GR 2 congress in Royaumont became my friends. I continued to see them at varying times or frequencies, depending on circumstances.

Lichnerowicz was a practicing Catholic. He went to Mass every Sunday. A good husband, after Mass he took his wife for a drink, alone with her. He respected the rules imposed in the past by the church: at a lunch we took together, I ordered steak and french fries. Lichné said, "It is Friday", studied the specialties on the menu and ordered a sole with tartar sauce. I said ironically, "I see that you mortify yourself." He replied curtly, "It is not a question of mortification, it is a question of discipline." However, his opinions seemed, to me, sometimes in little agreement with the dogma. I pointed it out to him one day, he answered, "I said that to the Pope and he approved."

A few years after the congress on General Relativity, Lichnerowicz organized, again in the Royaumont abbey, a colloquium on the philosophy of science with the collaboration of a young physicist, Pierre Aigrain. He was also a practicing

Catholic who entered "école navale" (navy school) in 1942 and remained unemployed until the liberation of France. Sent then to the USA by the Navy for naval pilot training, Aigrain went to the Carnegie Institute and came back to France with a PhD in electrical engineering to work with Yves Rocard in the physics laboratory of the ENS. Aigrain played a major role in the creation and development of electronics and computer science as we know them. Despite their investment in science and their religious conformity, Lichnerowicz and Aigrain were highly educated men in various fields. They both had particularly brilliant and quick minds. I have not kept memories of what was said in Royaumont, but I remember being dazzled by the public dialogues of the organizers.

The Sorbonne

I taught in Reims for only one year. At the end of 1959, Janet, holder after Chazy of the Chair of Celestial Mechanics in the Paris faculty of sciences, retired. Lichnerowicz suggested to me I apply, if not to the chair, at least to the position. His other student, Yves Thiry, expressed the wish to also apply. We were thus two candidates for this chair. Lichnerowicz supported my candidacy in private, but refused to make a public choice, fearing, he told me, to harm the future of Thiry. Fortunately Leray, though a friend of Thiry's father, had no such scruples. Both were already professors in the Colllège de France and therefore did not participate in the vote of the faculty, but their opinion had great influence. At the time, mathematicians constituted a commission which gave its opinion for an election in their discipline, but had no power of decision. Professors of all scientific disciplines were the assembly which elected the future holders of professorships. It was customary for candidates to visit these voters to defend their merits. I strongly wished to teach in Paris, and these visits were, for me, rather unpleasant except those to mathematicians, because I knew them already. I was then in the process of divorcing and the botanist Plantefol told me he would vote for Thiry, who at least had morality. It is true that during German occupation Plantefol was for Pétain and had no sympathy for my father. Danjon, a renowned astronomer and the director of the observatory, received me fairly well, exposing to me his own work on the astrolabe he had built and which had permitted a refined study of the motion of the Earth and some other heavenly bodies. He showed me a curve representing the results he had obtained, mostly a very regular curve, though disturbed by some peaks. Danjon told me these exceptional results should not be taken into account. I understood later it is common practice among experimenters, generally justified by a default in the experiment in question. Danjon did not seem shocked by my surprise. He probably would have preferred a researcher more involved in experiments, but Thiry was no more an experimenter than I was; besides, it was a Chair in mathematics. I keep an amused but good memory of my visit to Yves Rocard, then Director of the physics laboratory in the ENS. He told me, "I will follow the commission vote." It was good news for me. He added in a comforting tone, "But don't worry, we take anyone now, you will arrive in Paris." I think Yves Rocard was a great scientist. His non-election in the French Académie des Sciences is an example of the narrow-mindedness of some who are afraid of an idea which disagrees with beliefs forged by their predecessors; in the case of Rocard, the detection of water by dowsers. Someone told me that Darmois, then retired, supported Thiry — it surprised and grieved me. I called Darmois, he told me at once, before I asked him anything, that he admired the works I had done since my thesis and he did not doubt my election. Another proof that you must never believe what someone says someone else has said. I was elected with a comfortable majority. Though the privilege of the Sorbonne with respect to provincial universities was suppressed the year of my appointment in Paris, (without, I think, a cause to effect relation!), my mother was happy and proud of my election. She announced it to Professor Chazy's widow with whom she was on friendly terms. Mme Chazy did not want to believe that a woman had replaced the great man who had been her husband, "Yvonne in the Chair of Celestial Mechanics, it is not possible!" Thiry was elected in Paris a few years later. He wrote to me saying that Danjon wished that I share with him the teaching of celestial mechanics. I answered that this half-yearly teaching of three weekly hours could not be divided. Thiry did not keep a grudge towards me. I remained on good terms with him and his wife, a nice woman whom I met in the yearly colloquiums, the "journées relativistes".

I was candidate to a Chair which belonged to the mathematics department and only involved one semester with three weekly teaching hours. I had told Henri Cartan, President of the mathematics department that if elected, I would accept the completion of my service by teaching in the newly created faculty in Orsay. I would have done so willingly. In fact, the mathematicians working in Orsay did not wish the collaboration I offered them after my election. I did not try to understand their motives, but contented myself happily with the semester with three weekly hours, completed soon after with a section where I taught mathematical methods of physics in Paris.

During the first years after my election to the Celestial Mechanics Chair, I devoted myself to study and expose the beautiful and difficult mathematical works in this discipline of Tisserand and the brilliant Henri Poincaré. I took great interest in it, but did not find matter for a new discovery. Subsequently, the use of computers allowed for remarkable progress in the numerical determination of the motion of celestial bodies, especially of artificial satellites, the launch and usefulness of which have not ceased to increase. The mathematicians then gave teaching hours to a specialist of the subject, Jean Kovalevski, a former student of Danjon who had done a PhD in Yale. He is now a member of the Académie des Sciences in the "Sciences de l'Univers" section. I appreciated his collaboration, and it allowed me to teach subjects nearer to my research field, mathematical problems arising in fundamental theories of physics. This teaching led me to write, a few years later, a book, "Géométrie Différentielle et Systèmes Extérieurs", that was published by Dunod in 1968, now out of print.

Marriage and Son

Gustave had asked for his divorce for the wrongs of his wife Marie already in 1955, requesting the custody of their three children. The divorce took place only six years later under reciprocal wrongs. The issue of custody was not the same: the eldest boy had come of age and the second one was very close to it. Marie wanted, for reasons of finances and self-respect, that Gustave take all the wrongs in the divorce; Gustave refused for the same reasons. She even went to the Supreme Court when she lost the case on appeal. Our two divorces at last pronounced, my marriage with Gustave Choquet took place on 16 May 1961 in Sceaux where he was domiciled. The mayor of Sceaux gave some decorum to the ceremony in a pleasant room of the town hall, in the presence of my mother, my brother, my daughter Michelle and our witnesses, Adeline for me and for Gustave, his former master Arnaud Denjoy. We had the pleasure to have the presence of our friend Luc Gauthier who, also living in Sceaux, had seen a displayed announcement of this marriage. We went to the wedding feast on the terrace of a restaurant with a beautiful view in the nearby Plessis Robinson. Denjoy courted Adeline, who drew from it some pleasure despite his age.

The university holidays arrived soon after the marriage. Gustave, much attached to his children — one of the qualities which had attracted me — had rented in Vallorcine, in the Chamonix valley, a cottage near one of his friends, Revuz. We spent July there with my stepsons Bernard and Christian, big boys with whom I always had excellent relations, and with my daughter Michelle. To keep her company, we had taken, with us, her friend from the Résidence de Tourvoie, Marie-Françoise, about fourteen years old and already a good little housekeeper. The boys and girls formed two teams who took turns with good will to wash dishes or other small tasks. Gustave and his sons made some climbs with the Revuz. We took together some lovely walks.

I had the great happiness, soon after our marriage and our arrival in Vallorcine, to conceive the son I had longed for. This happy event took place in a beautiful meadow in altitude during a walk with Gustave that we took alone for once. My pediatrician and friend Lepintre, when I told him about my remarriage and that Michelle wished I had another child, answered that she would perhaps be less pleased when he would be there. In fact Michelle was very happy to have a little brother and chose his first name, Daniel. She remains very attached to him despite the differences in their choice of which values are important in life. Daniel was given middle names from those of his grandfathers, Georges and Gustave. The name Georges seemed to me required for my son, and his father wished to add the one of his own father. It had been given to him because he was born during the war in 1915. His pregnant mother had had to flee the invaded north with her one-year-old daughter. Without news from her soldier husband, she had believed him dead and had given his name to the baby. In fact, my father-in-law-to-be had not died, but had been a prisoner in Turkey. He fortunately rejoined his family at the end of the war; there were then two Gustaves.

Claire, the daughter of Gustave, then thirteen, did not come with us to the Alps, preferring, to my regret, to spend that month with the scouts and stay with her father in August. Claire liked the sea. We reserved, for August, an apartment in Saint Cyr, from where the waterfront, beach or rocks, is easily reached on foot. Unlike her brothers, Claire had a grudge towards me for having married her father. Without being positively unpleasant, which would have gained her paternal reproaches, she was so by her whole attitude. She always behaved as if I did not exist. I understood her and felt no resentment. I also understood Gustave, who tried, in every circumstance, to please his daughter, without worrying about what would have pleased his wife, but still, it was painful to me. The behavior of Claire towards me has completely changed thereafter. She even wrote to me, recently, a letter apologizing for her wickedness towards me at that time. I answered that she exaggerated her faults. I had understood and never kept a grudge towards her.

Life at "allée de Trévise"

After the 1961 holidays we lived briefly in my pavilion in Fresnes, but we soon moved in a more spacious apartment, on allée de Trévise in Sceaux, reserved by Gustave when it had been still under construction. This apartment was on the ground floor, but for me it was not a fault, for I like to have my feet on the earth. There were four bedrooms, rather small: one for the parents, one for Michelle and one for the expected baby. The last one was occupied for a time by my stepson Christian, then by the family help we hired after Daniel's birth. We had the pleasant surprise to find, in the same residence, Henri and Madeleine Cabannes who had booked an apartment when Henri was appointed in 1960 to a Chair of Mechanics in Paris.

My pregnancy went on without problem, as well as the delivery in a pleasant clinic near the Bois de Boulogne. Daniel was a cute baby. He weighed only 2 kg 700 g, which worried me a bit. My friend, Dr. Lepintre, took the trouble to come from his place, at rue Duguay-Trouin, to examine him. He reassured me on the health of my baby, adding: "2 kg 700 g is better than 2 kg 500 g".

We already had a pleasant and efficient housekeeper aged about forty, Mme Wimel; however, working in the Protestant temple of Sceaux, she could come only two or three hours a week. Day nurseries where one leaves babies did not exist yet. Thanks to a professor in the lycée Marie Curie, I had the good luck to find, quickly, a family help to take care of a number of household issues and of Daniel when his parents were both working in Paris. She was a young, cheerful and active girl named Ghislaine. She manifested, at once, much sympathy to Michelle, who returned her affection. We had all confidence in her regarding Daniel; however, Gustave and I only abandoned when necessary the pleasure of taking care of our baby. Our shared love for our children was a powerful link between us, which I had not found with Léonce.

In the beginning of the summer of 1962, Gustave left us to participate in the International Congress of Mathematicians in Warsaw. He very much wanted not to miss this congress, especially because he kept fond memories from his stay in Poland as a visiting professor just after the war. He was delighted to meet friends again. We rented, for the summer, a villa in Hossegor. We went there first: me, Ghislaine, Daniel and Michelle. After the return of Gustave, Michelle was replaced by the second son of Gustave, Christian. These holidays did not leave me special memories, except that in the absence of Gustave I drove the DS, which was not possible for me in his presence.

Gustave and I went alone with baby Daniel to spend the Christmas holidays of 1962 in a hotel of Roquebrune. The aunt of Adeline, who lived in that village, knew the owner. We were received like friends, seeing the arrival of champagne and delicacies on Christmas Eve in the room where we were sleeping. The weather was splendid and we made beautiful excursions around, with our DS or on foot — once, more on foot than we wanted, because Gustave had slammed the door when we left the car, with the key inside. This incident does not prevent me from keeping a good memory of these Christmas holidays under the southern sun.

Upon our return, life, teaching and research resumed their courses. Gustave's children came to see us on week-ends and Ghislaine continued to help us.

Work in 1962–1963

My pregnancy was without problem; however, my mental faculties were a bit asleep, which is not compatible with advanced research. I used my free time to write my first book, "Mathematical Problems for Physicists", a complement to a recent book by Lichnerowicz, "Algèbre et Analyse Linéaire". Both were published by Masson, where had appeared the books of my father. My book was translated in English and published in the USA by Holden Day.

In 1963, I studied the global in time properties of the "propagators" associated to the kernels used in my thesis for the solution of the Cauchy problem. These propagators were used by Bryce DeWitt and Lichnerowicz to formulate brackets corresponding to the quantization of a gravitational field on a given space-time. My work was published in an Italian journal whose editor, Enrico Bompiani, was known well by Lichnerowicz. I had obtained, I thought, interesting results. At the time they did not have significant applications, but some of the properties published in French in this Italian paper reappear in English in recent works on gravitational waves.

My First Stay in Les Houches' Summer School

In the summer of 1963, I was invited to participate in the school of theoretical physics created and directed in Les Houches by Cécile DeWitt. I did not give a course, but I was a kind of privileged participant, with room and board provided for me and my family, that is, in this case, my husband Gustave and my son Daniel, then aged a little more than a year and already very lively. This caused a problem: the cleaning woman left lying on the ground a pot of rat poison flavored with chocolate, and Daniel tasted it. Although he did not have time to consume much, we took him to the hospital where he was given a gastric lavage. The doctor decided to keep him for the night with me to watch over him. Daniel was in an excellent mood, and I was not really worried. The cleaning woman was blamed, and life followed its course, interesting and pleasant. The school in Les Houches is located in a beautiful site surrounded by nature above the village which bears this name in the Chamonix valley. The 1963 session was named "Relativity, Groups and Topology". The lecturers were physicists or mathematicians of value, of different ages, with most of them very friendly. Gustave, although a pure mathematician, had an open mind and liked to talk with specialists of other fields. I remember particularly having appreciated, during this stay, the presence of Roger Penrose, a devoted young dad and a scientist full of original ideas. I also appreciated the young Ray Sachs, who had just obtained interesting results on gravitational radiation, which earned him a professorship in Berkeley, a great Californian university where Abraham Taub was a professor (refer later to my visits in this university). Unfortunately Sachs has been too disappointed by his failure to generalize enough his first results. He gave up research to devote himself to teaching. Ray is not the only example of a scientist discouraged from the pursuit of research by obtaining very young an exceptional but incomplete result on a difficult problem. Penrose, on the other hand, has become famous, ennobled by Queen Elizabeth. His visionary conjectures continue to give work to many theoretical physicists. He has written several fascinating books for the general public, speaking of physics and philosophy. In spite of his successes, he remains a man without pretension, open and friendly. He came to listen, a few years ago, to a talk I gave at a seminar in Oxford. Afterwards we had lunch with a few colleagues and the conversation turned to the publication of his complete works. Penrose said, "My problem is to know if I must correct my mistakes before publication." It is a great quality to accept to recognize a mistake, even small. Few human beings, scientists or not, are ready to do it. I have seen Roger again, always with pleasure, in various meetings, the last one being in Berlin in 2015. Since then I bought Penrose's book of more than a thousand pages with the appealing title "A Road to Reality". It is an extraordinary achievement.

Visit in Cornell

While I was expecting Daniel, Gustave had a few weeks' stay in the USA where I could not accompany him. He received an invitation to spend the three months of fall 1963 in Cornell, one of the most important American universities, to do research in collaboration with Professor Carl Herz. Cornell is located near the small town of Ithaca, in New York State; the natural scenery around it is beautiful. Carl wrote that the university wished to invite me also to give an "undergraduate honors course" of mathematical analysis. Such courses are for undergraduate students selected for their abilities. Gustave wished to accept. I wished it also, on the condition that we take, with us, Daniel of course, but also Michelle. Ghislaine, after some hesitation, accepted to come with us, solving thus the problem of Daniel. Gustave suggested we leave Michelle in France, but I refused. Her father made some difficulties, asking the court to order me to put my daughter in a pension in France during my stay in the US, taking the pretext of her schooling and the fact that he would not be able to see his daughter as usual for Christmas, our return being only in January. Thanks to a letter from the lycée Marie Curie, the court allowed me to take Michelle. I enrolled her in Ithaca in an American school, in a grade corresponding to the one she would have entered in France, with French as her foreign language class and no Latin. Michelle had always been bored with the study of Latin and wished to abandon it. In addition, she followed correspondence courses. Her beginning at school was a bit difficult, but within a month she managed English well enough to understand. It took her longer to decide to speak, though she was once classed first in English grammar! It is true that grammar was then better studied in France than in the USA. I am afraid that emails will soon finish drowning grammar everywhere.

We had rented a house for our stay in Cornell, but when we arrived in September, the owners still lived there. The rental agency had arranged, for us, a stay in a villa located on the edge of a lake until the owners' departure. That villa was pleasant, in a beautiful landscape, but it was a summer house with windows devoid of glass. We were quite well in it. Fortunately, we moved into the house we had rented in town as soon as it was free because the very next day, the weather, sunny and mild so far, suddenly cooled — snow covered the yard when we woke up. America is a young country and even nature there is violent! In fact the surroundings, all white with snow, were magnificent. I have good memories of our stay in Cornell in spite of the lack of contacts with colleagues sharing my mathematical interests, or having spouses whom I would have befriended. I remember one of these ladies telling me one day, "One cannot be a good mother and a mathematician." I asked, "Why?". Answer: "Because the child must come first". My repartee, "And how do you do when you have several?" stopped the dialogue.

Ghislaine, a smart girl, had befriended a French woman, a supermarket assistant, who was married to an American man. Ghislaine told us that this French woman, though she was not an intellectual, thought Americans shamefully ignorant. Indeed the cultural level of the average French person is, or at least was sixty years ago, much higher than that of the average American.

The level of the undergraduate students, though "honor", that is, among the best, was analogous to the French students of "licence" (bachelor level) of the time. My classes of about thirty students did not cause me any problem. It is traditional in the US to ask students for an evaluation of their professor. The reports I got after my courses were all favorable. Their only reproach concerned the book which had been recommended to them, as is traditional in the US, but the choice of this book came from the previous teacher.

During our stay in Cornell we were absent two or three days, each of us giving a lecture in Cambridge, a suburb of Boston where are located the famous universities of Harvard and of MIT. We left Daniel to Ghislaine and Michelle, both of whom took good care of him. These girls also wished to travel, to see something else of the States other than Cornell and Ithaca. Michelle was now doing well in English; Ghislaine was twenty-one, intelligent and responsible. We organized, for them, a trip to New York, with a little stay in the YWCA (the Young Women's Christian Association — I myself stayed there several times), then an excursion to the Niagara falls. Gustave and I had both seen them, separately, before. These young ladies managed very well and came back quite happy with their small holidays.

Beginnings of Relativistic Astrophysics, Texas Symposium

Gustave left briefly to give a lecture I don't remember where. I myself escaped a few days to participate in the first international congress on relativistic astrophysics in Dallas. In his article "The first Texas symposium", published in Physics Today and reproduced on the web, Englebert Schücking says that the term "relativistic astrophysics" was invented by him and Alfred Schild when they organized in Austin a congress on General Relativity in 1963. Schild, born in 1921, had spent his childhood in Turkey before the settling in Vienna (Austria) of his family. This stay had given him an immunity against various bacteria native to southern countries, which make life difficult for human beings too hygienically brought up, for instance in the USA. Schild had started his university studies in England, interrupted at the beginning of the war because of his German passport which had had him imprisoned. Released and sent to Canada, he could continue his studies and work on a thesis under the direction of Leopold Infeld, known for his collaboration with Einstein on General Relativity. When I knew Alfred Schild he was a professor at Austin University, married to a very pleasant American-born woman, Winifred, often called Winnie. Alfred was a sociable man, very kind. He had known how to attract to Austin talented relativists of pleasant character: Englebert Schücking, Roger Penrose, Wolfgang Rindler. Other renowned relativists, among them Luis Bel, Roy Kerr and Ray Sachs, made long sojourns there. Schild brought to Texas, but to Dallas because Austin was full, a very active relativist, Ivor Robinson, who continued there his career. I knew him well because he had befriended Lichnerowicz and had come several times to Paris. Shücking writes, in his article on the first Texas symposium, that Schild and he, whirlpool of activity, had wanted to reunite other relativists, taking advantage of the oil wealth of Texas. There was much talk at that time about quasars, mysterious celestial objects that had been recently discovered. Englebert and Alfred obtained important funds from oil tycoons for the congress they were planning by calling it "Texas Symposium on Relativistic Astrophysics". Shücking says he is the inventor of the last two terms. He adds that these funds were so generous that they invited all the relativists whose names occurred to them. I was among those, and I went to Dallas with pleasure, then to Austin in Schild's car. I don't remember the content of the lectures given at this colloquium, not even mine. I remember only that Alfred had invited us to an excellent lunch and that he had dozed during the presentation I made just after. I had been especially upset because we had become friends from our first meeting. I thought this should have helped him to stay awake. However I did not keep a grudge and I remained on excellent terms with him and his wife. I saw them with pleasure at other Texas symposiums where I was invited, for instance, in Dallas again in 1968, then in 1972 in New York where Englebert had become a professor in the Courant institute. Schild unfortunately died from an infarctus in 1977; he was only fifty-five. I did not meet Winnie again: it is a bad side of the marriage with a lecturer of congresses, wives are not invited after the death of their spouse. On the other hand, I had, subsequently, the pleasure of seeing Englebert several times. I even slept at his place in New York when I went there in 1981 on the occasion of a congress in the honor of the famous mathematician K. O. Friedrichs. I had not worked in collaboration with Englebert, but we had common scientific interests. I appreciated his understanding of human issues and his humor very much. I wrote an article for the proceedings of the colloquium organized at his retirement.

Relativistic astrophysics has enjoyed a great development with the considerable improvement of telescopes and the launch of satellites which multiply the information we have on the cosmos. The Texas symposiums, which seldom take place in Texas since the first few but have kept the name given by their founders, are in their twenty-eighth edition which will be held in Geneva. The twenty-seventh one, celebrating the fiftieth anniversary, took place in Dallas like the first one. The congress now has many sections: mathematics, numeric, observation. The participants, unlike those of the first congress, are mostly relativistic astrophysicists, because relativity is now essential in astrophysics.

The Relativity Center in Austin

Despite the departure to other universities of the famous relativists quoted above, the General Relativity Center, created in

Austin by Albert Shild, did not cease to exist. Bryce DeWitt and his wife Cécile both left Chapel Hill and moved to Austin — Bryce to supervise relativists, with Cécile as a professor in the astronomy department. Austin University, wishing to enhance itself with the presence of the DeWitt couple, had found this solution to circumvent the anti-nepotism law which forbade husband and wife to be professors in the same department. Until the abolition of this law, Cécile, a courageous and conscientious woman, learned and taught astronomy without trouble, just as I had taught celestial mechanics.

After the installation in Austin of the DeWitt couple, I went there several times to work on the books written in collaboration with Cécile and their reissues. I also appreciated during my visits in Austin meeting Bryce, the husband of Cécile. He was a great physicist, an original mind, independent and creative. He was also a man of character, daring, and loving the great outdoors. He made ascents in the Himalayas, and flew over the Sahara (he was an experienced pilot) looking for a famous meteorite. Cécile helped in organizing expeditions and joined him as much as she could. I did nothing as spectacular as the DeWitt family, which I regret a bit. But I keep an excellent memory of some expeditions I made with them, for instance sleeping under the stars in Padre Island, a desert at the time.

Other members of the "Center for Relativity" in Austin I knew well are Richard Matzner, who was responsible for the "Great Challenge" created by NASA for the determination by computer of n-bodies motion in general relativity, and Larry Shepley, who had the kindness to collaborate with Cécile and I to get us a grant for our work when I visited Austin in 2003. I had the pleasure on this occasion to meet Karen Uhlenbeck again, a great American mathematician and a feminist militant. Last and not least, J. A. Wheeler himself went to continue his activity at the Center for Relativity in Austin when he retired from Princeton in 1976. He stayed there for ten years. I have been happy to discuss science with him and also to have lunch at his home and chat with his wife Janet.

Gravitational Waves

In General Relativity, a gravitational wave is a perturbation of an ambient gravitational field. The resulting field is an approximate solution of the Einstein equations. The words "perturbation" and "approximate" are not mathematically precise without further definitions. There are several kinds of construction of what is called "gravitational waves". The properties of these waves were, in 1963, a largely open problem. It was satisfying to have some precise mathematical results with well-defined approximations. Einstein himself had proven a particular result: a perturbation of the special relativity metric solution of the classical wave equation satisfies, in first approximation, the Einstein equations in vacuum, if this perturbation satisfies, also, four linear equations that can be considered to define the wave polarization. This result was interesting, but very partial, considering only the neighborhood of Minkowski space-time. Moreover, it did not justify the splitting of the equations into wave equations and polarization conditions, nor did it give an estimate of the error due to the approximation.

I started to work in Cornell on high frequency gravitational waves by using the general asymptotic expansions introduced by Jean Leray in his course at the Collège de France. I had been very interested during the years 1961–1962, in this construction of high frequency waves as asymptotic solutions of general systems of linear partial differential equations. The method used by Leray generalized, in an essential way, the WKB (Wentzel, Kramers, Brillouin) method, itself generalized by Peter Lax without, however, becoming applicable to non-linear equations. Although Leray himself has not been interested in that problem, the expansions he introduced can be applied to non-linear equations. In the work begun in Cornell and continued after my return to France, I used the general expansions considered by Leray to study high frequency wave solutions of the Einstein equations, oscillatory perturbations of very short period of a general relativistic space-time called background field. In these first works I constructed, with precise approximation, gravitational waves where perturbation and background field contribute equally. I showed that they propagate like light and necessarily satisfy four polarization conditions that I wrote. This work was the subject of a paper in the "Comptes rendus de l'Académie des Sciences", and the thesis that Antonio Ribeiro Gomes had come from Portugal to do in Paris. He had first posed a subject to Lichnerowicz who, already overloaded with students, had sent him to me. Gomes was a serious worker. He had developed, following my guidelines, the work I had sketched. I regretted that he left as soon as he obtained the first results to defend a thesis in Portugal and take a professorship which happened to be vacant in Coimbra University. He was active thereafter in developing the mathematics department, and became its director. I was happy to visit him several times in this old and pretty city along a river where, at that time, women still came to rinse their laundry. Probably thanks to him, I was invited to Lisbonne, a very beautiful town where I had happy stays.

Holidays in Corsica

We went back to France during the winter 1963. During the holidays that Michelle spent with her father and Ghislaine with her family, Gustave and I decided to go, with Daniel of course, to

Corsica, a place neither of us knew and where we hoped to find sunshine. We arrived by plane, rented a car and left for adventure. Gustave wished to see mountains, therefore we went towards the interior and its splendid landscapes. We had no trouble finding rooms in hotels on our way, but they were not heated. Fortunately, blankets and eiderdown were not missing and neither little Daniel nor his parents got a cold. The mountain roads were covered with snow. I remember a descent which I adjured Gustave not to take, having seen an overturned car in the ditch that ran along it. But Gustave was a fearless man who followed no advice, he entered it. I demanded that he stop and went down on foot, carrying Daniel. I was not afraid for myself, but I did not want my son to run a risk. We arrived safe and sound at the bottom of the slope, so did also Gustave and the car. I don't keep a bad memory of these holidays. Corsica is a beautiful country, but I do not advise anyone to drive around there in winter.

Return to Sceaux

When we were back to Sceaux in the beginning of 1964, the lycée Marie Curie refused to admit Michelle in third. The headmistress wanted her to do a fourth again. Michelle absolutely refused. I then made her continue the third at home, helped by some private lessons. At the end of the school year Michelle was accepted without problem at the entrance examination in modern (i.e. without Latin) second. She was a very good student. Mme Leray told me, "Michelle triumphs in a class of bad students." It was true.

A painful incident for Ghislaine was when the doctor and surgeon mistook menstrual pains for appendicitis. In our absence, Ghislaine had abdominal pain so severe that she told Madeleine, who worried and called a local doctor. He then diagnosed it as appendicitis and had the victim transported at once to the hospital where she was operated on in emergency. The surgeon told me later that she had exaggerated her symptoms. In fact, I am sure that she did not have appendicitis, because she had the same pain her following menstruation, not an unusual phenomenon in girls of that age. It was the second case I had come across of an appendix uselessly removed. Fortunately it does not seem that its absence causes problems.

In the beginning of 1964, I found myself pregnant again, happy to give a brother or sister to Daniel. We were both invited to a congress of a few days, in Taormina, Sicily. Ghislaine encouraged us to accept, assuring us that she would take over our household duties very well. We thus left, reassured by the joint presence of Ghislaine, Michelle and Mme Wimel. Back in Sceaux, we found, in top form, those we had abandoned briefly. Unfortunately, I had a kind of flu and, perhaps because of that, I lost the baby, a small embryo of about two centimeters.

Seattle

After my miscarriage, we decided that I would accompany Gustave to Seattle where he had accepted an invitation for July 1964. I had given up on this trip because of my pregnancy. Michelle spent that month with her father in Chamonix and Ghislaine took her holidays. We rented a small house in the suburbs of Seattle. I was a housewife. Nature is very beautiful in the neighborhood of Seattle. On weekends we took pleasant walks with Daniel, now a very charming blond little boy of two. Gustave spent the weekdays in the university and Daniel and I remained alone. Fortunately the weather was very nice, and the neighbors had children who became Daniel's playmates. In the United States, the yards are not closed between neighbors. I could thus indulge in my favorite occupation, mathematical research, a few hours a day, while keeping an eye on my offspring. Leray had taken anew, in collaboration with the mathematicians Garding (Swedish) and Kotake (Japanese), his work generalizing, in an essential way, that of Peter Lax, himself inspired from the method called WKB. Garding, Kotake and Leray published, in 1964, a long article, "Uniformisation et Développement Asymptotique de la Solution du Problème de Cauchy Linéaire à Données Holomorphes, Analogie avec la Théorie des Ondes Asymptotiques et Approchées". During my stay in Seattle, I got particularly interested in the aspect of their work on the properties of solutions in the complex domain. I succeeded, I must say to the surprise of Leray, in extending, in a relatively simple manner, these properties to non-linear equations. Garding, whom I met later, told me he had given a talk on my results in his seminar in Lund. All mathematicians know the pleasure brought by the proof of a new theorem. My work in Seattle leaves me a pleasant memory.

Good times, not related to mathematics, included the admiration of splendid landscapes during our trip to Vancouver where Gustave had been invited to give a talk. One day, as we were walking, stopping from time to time to leisurely enjoy a place, we were surprised to see a man following us. Finally he told us his concern: the municipality gave, to whom wished it, a piece of land of the person's choice. He had not yet chosen and he was afraid that ours was the one he would want himself. He was relieved to learn that we did not aspire to be land owners in Canada. We would probably not have been eligible for the British Colombia administration liberality but I sometimes regret that we did not try.

Return to France

I found myself pregnant again rather quickly, too quickly after a miscarriage, I was later told. As soon as I was back in Paris, I went to see the gynecologist who had helped me for the birth of Daniel, because I very much wanted to have another child. The doctor asked me to do a hormonal analysis, found it was not quite satisfactory and ordered a medication, perhaps Distilbene which has since been accused of long-term adverse side effects. In fact, fairly fast, in spite of the medication, my hormone level collapsed. The doctor told me to stop the drug; the fetus was probably dead. She said it was normal for my age, reminded me that my previous pregnancy had finished with a miscarriage, and added that the last baby I had had been smaller than normal. The decision to stop the treatment sounded, to me, like a death sentence difficult to accept. I consulted my friend Dr. Lepintre who gave me a recommendation for an eminent gynecologist he knew. This specialist received us quickly, read my file and told me to continue the treatment. However I understood, from his eyes when he shook my hand at my departure, that I would not have this child. I started a few days later to have contractions and I went back to see the gynecologist as he had recommended to me. He decided to do a curettage; there was a remainder of placenta, but no embryo.

Christmas in the Balearic Islands

Gustave, Daniel and I spent the Christmas vacations in a hotel in the Balearic Islands. We found there beautiful sunshine and the mild temperatures we had missed in Corsica. We rented a car to go around the island. By chance, Laurent and Marie-Hélène Schwartz had also come to spend these holidays there. Laurent and Gustave were buddies, of the same year in École Normale. Marie-Hélène was very nice with the young Daniel, apparently happy of his company. Indeed I believe that the Schwartz were happy to share our four wheels because neither of them drove. They did not hide their disappointment when they joined us the last afternoon and learned we had returned the car, intending to walk.

11

FIRST YEARS IN ANTONY 1965–1968

Moving to Antony 1965

As both of us loved nature and hoped to see the family grow, Gustave and I decided to leave our apartment for a villa, also located near the Parc de Sceaux. We chose a house located on the border between Antony and Sceaux, also near Bourg la Reine. This house was big enough for each child to have a room and the parents to have a bedroom and an office. It was very near to a subway station (RER B) so it was easy to reach the Henri Poincaré Institute where the parents worked, and very near to the elementary school in Sceaux, not too far from the lycée Marie Curie for Michelle and later Lakanal for Daniel. We moved to this house in 1965.

Ghislaine had left us as she was getting married. She suggested that we replace her by asking the sister of the husband of a friend of hers to come from Martinique. We would give her an advance for the fare and take it back on her salary. We trusted Ghislaine exceedingly, and we had met and liked her friend, thus we accepted this proposal. We did not regret it. Agnès was hard-working, a good cook, and had good presentation at the table when we had guests. We had never been better served.

In summer 1965, we rented a small house in "île grande" on the north coast of Brittany, so that my stepdaughter Claire, who was staying with the Scouts in a nearby village, Perros Guirec, could join us easily. I was happy to see, again, the beautiful Breton countryside. We had very nice weather, unusual, they say. Agnès went with us for walks. She taught us to eat freshly caught raw shrimps. She seemed less hurt than myself when a foul-mouthed man who passed by on the trail we followed on the beach front said, "She is tanned, that one." Unfortunately, at the time the quality of contraception was not at its top, and Agnès had warm blood. A few months after our return to Antony, she was ill and asked me to bring a doctor whose contact she gave me. This doctor came to her room, and let me understand she had tried an abortion, illegal at the time. He added that her pregnancy would follow a normal course.

Myself, I did not wish to make a third try for maternity. Daniel had grown, and Michelle had become a teenager, not interested anymore in a baby from her mother. I had seen that Gustave, like most men of his generation, was not for gender equality with regard to domestic or parental duties, and authority. When the law passed on shared authority he said, "Shared authority is no longer authority." However, women were not yet, at that time, masters of their maternity, and I found myself pregnant again. In fact, I immediately wished to lead, to its term, the little life germinating in me.

My Daughter Geneviève

I quickly went to see the specialist recommended by Dr. Lepintre, the hormonal analyses were found excellent and Geneviève remained quietly during the nine months in the maternal womb. Perhaps, the curettage had refreshed my uterus. Her pregnancy worried Agnès. I told her I was also pregnant and she and her baby could stay home with us, she would help me take care of mine. Agnès accepted this proposal with relief. Welfare took care of organizing her delivery and found her lodging for a few weeks before and after the birth.

In July 1966, Gustave went to a congress abroad, while I remained alone with Daniel and Michelle. The latter, intelligent, had good results at school, but her independent spirit caused me worries. I had rented, in Les Lecques, an apartment for our holidays but it was not free yet. I went, however, to Les Lecques with my two children and Michelle's cat that she had refused to part with. I took pension in a small hotel near the villa Dar El Chott where I had stayed a few years before. The hotelier, very understanding, had given Michelle and her cat a room on the ground floor adjoining the hotel. I have good memories of this worriless period. The room I shared with Daniel had a low window which opened right on the beach. I worked sometimes towards the completion of my book "Géométrie Différentielle et Systèmes Extérieurs" while watching my son making sand castles. A friend of Michelle, Françoise R. joined her impromptu. I wondered how to accommodate her, when the owner of the apartment where we were to stay a little later told me that the lodging was free and encouraged me to come at once, with the three youngsters. We did that. Michelle and Françoise gave me some worry, having wanted to go wandering. These fifteen-year-old girls were brought back to me the same day by the police. A few days later, Michelle left as planned to spend the second part of her holidays in Chamonix with her father at the Defaixes. Gustave arrived and the holidays continued pleasantly. The very nice and efficient Mlle Massel, who had helped me in the past, had accepted to do so again a few hours a week.

At the end of August we returned to Antony. Agnès came

back with her newborn son, Philippe, who was a beautiful big baby. We had arranged a large room for them on the ground level, with window, water and gas, and toilets nearby. It had been a kitchen and living room for the employees of the previous owner, a builder from whom we had bought the house. During the absence of Agnès I had convinced Mme Teilhet, who had worked for me in Fresnes, to do it again. She accepted out of kindness and continued to come to help Agnès, who was back with her baby. Mme Teilhet stopped working, upon her wish, only after Agnès left in 1968, and I hired a courageous Portuguese girl who was both my house-keeper and family help.

The Birth of My Daughter

The eminent gynecologist who had started following my pregnancy had a heart attack and took a long leave. I did not feel good with his replacement, and the clinic where he made his deliveries was not easy to reach from our home. I made a reservation in a clinic near Antony. In fact, the gynecologist attached to it did not trouble himself to come. I broke my waters in the afternoon of September 28 1966, and Gustave took me at once to the clinic. Around ten o'clock in the evening, nothing new had happened and the midwife on duty told me she was going to induce labor to avoid suffering for the baby. I agreed immediately, asking her only to tell Gustave, who I knew was very happy to be a father again and wanted to be present at the birth of our child. He soon arrived and the midwife installed a perfusion which provoked contractions so violent that I could not restrain my screams. The reaction of my husband was not compassion for me but to tell me, "Don't scream, you will hurt the baby." Fortunately, the birth came fairly fast and Gustave said with great satisfaction, "It is a girl." It is true that he had already three boys and only one girl. As for me, girl or boy, I was happy to have a beautiful baby, of reasonable weight, at 3 kg 350 g, which was more than my two other children's. I did not linger in the clinic and I started to breastfeed my daughter. Unfortunately I had caught, perhaps because of the drip, a viral hepatitis. I felt very poorly. Lepintre brought a specialist who confirmed his diagnosis and ordered a treatment by cortisone, a long rest in bed and interdiction to come near my baby. I am told now that cortisone was the wrong prescription, but I have been happy of it: painful symptoms disappeared fast and resting in bed was not an unpleasant experience. Of course I regretted being able to see my little girl only from the threshold of the room where I was lying down, but I knew she was well taken care of by Agnès, under her father's supervision. When, after a few weeks, I could take care of her myself, it was a great pleasure, because she was a baby who generously lavished smiles.

Children

Daniel and Geneviève both went to the primary school at allée de Trévise in Sceaux, though our domicile was now in Antony, avenue d'Alembert. There is, at the end of that street, a private school called "The New School". I had kept a bad memory of the "cours Gernez" of my childhood and it was contrary to my soul, that of a public school teacher, to put my children in a private school, even a secular one. Moreover, the few children I knew who were put in that school had been so due to special problems. Hence Daniel and Geneviève continued the twice-daily round-trip of a short kilometer from Antony to Sceaux. The entrance examination to sixth of my childhood had been removed. The first classes of secondary school had not yet been renamed "college". When he was ten years old, Daniel entered in sixth at the lycée Lakanal in Sceaux, easy to reach from our home. Girls now also admitted, Geneviève could also, at the same age, follow her studies at Lakanal, avoiding, thus, the fate of her older sister who had to endure the lycée Marie Curie. The Dean of Lakanal was an intelligent and open-minded man. Daniel was brilliant in sciences, less in letters, while Geneviève's performance was more even. Their secondary studies did not give us any worry. Both of them have kept, I believe, a good memory of their years in the lycée Lakanal, in contrary to Michelle's at the lycée Marie Curie.

"Battelle rencontres" 1967

The young ones, Daniel and Geneviève, gave me only happy times. They were a powerful link between their father and me. However, I was somewhat concerned about my oldest daughter Michelle. Intelligent, she had good academic results; she had obtained, without problem, the scientific baccalaureate in spite of her youth, at sixteen years of age. But she did not get along well with her stepfather, an authoritarian man unwilling to compromise, while Michelle was an adolescent with an independent and voluntary nature. My wish to change the atmosphere for her was an additional motive for me to accept the invitation I received from J. A. Wheeler for a stay of a month in Seattle where he was organizing, helped by Cécile DeWitt, "Battelle rencontres" between physicists and mathematicians.

The "Battelle memorial Institute" is a private, independent organization, financed by funds left in 1923 by a manufacturer, Gordon Battelle. According to the Internet, the activity of the Battelle Institute, now located in Ohio, has grown considerably and finances researches in various domains, mainly of science applied to industry. The Battelle rencontres to which I participated in 1967 were purely theoretical. The participants were accommodated in villas in Seattle, near the Institute building where the lectures took place in a room of relatively modest size. I found, in Seattle, a friend who had been a student in the ENS at the same time as me, René Thom, who had proved to be an outstanding mathematician as Henri Cartan had predicted. Thom, founder of the mathematical catastrophe theory, was also a lover of philosophy and I enjoyed chatting with him. I met, at these Batelle rencontres, other great geometers: Raoul Bott, Sigurdur Helgason and a few more. Lichnerowicz was also there.

Among relativists, there was Roger Penrose whom I had met in Les Houches a few years before, now already famous for his visionary conjectures. There was also Robert Geroch, a lucid and imaginative mind, with whom I had stimulating discussions. Stephen Hawking, Steve to his friends, who had since become a world celebrity, participated in these rencontres. He had been present in Seattle with his first wife and their young baby. He suffered from the first attacks of his terrible disease, the syndrome called Charcot in France, Lou Gehrig in England, walking with crutches but not yet having difficulty in speaking. I remember interesting discussions on global properties of Einsteinian space-times and also pleasant conversations during excursions with this witty young man.

Research: Global Hyperbolicity

I think that the Batelle rencontres of 1967 are the first long meeting of the best in their domains, pure mathematicians with theoreticians of General Relativity. These encounters have opened the way to global existence problems, that is for an infinite time, for the solutions of Einstein equations. They also have stimulated research on the formation of singularities or simply end of causality.

During my stay in Seattle, I worked at applying, to the Einstein equations, the definition given by Leray of global hyperbolicity for general systems of partial differential equations. The geometric interpretation is easier when the unknown is a Lorentzian metric. I reformulated the Leray definition in simple geometrical terms for the space-time metric, and I obtained an easy-to-check criterion. I proved rigorously a few theorems, and studied their link with previous, sometimes heuristic, results obtained by Penrose, Hawking and Geroch. I gave a small course on the subject and wrote a pedagogical chapter for the book "Battelle Rencontres 1967" edited by C. DeWitt and J. A. Wheeler; published by Benjamin in 1968. I had the pleasure to see it used and quoted in two meetings where I went many years later, by Anna Maria Candela in 2006 in Bari and by the Spanish man Sanchez in Berlin in 2009.

Daily Life in Seattle 1967

Life was pleasant during these rencontres. Proximity but diversity of knowledge and research subjects of participants made scientific discussions very interesting. The comfort of the facilities and the beauty of the site favored a warm atmosphere. The administration had provided a car for each family. Lichnerowicz was there with his wife, but neither of them drove. I was, therefore, led to serve as their driver for excursions in the area. I thus got to know Suza Lichnerowicz better, an intelligent woman, who was very cultured and possessed clear judgment.

Michelle, I think, was happy in Seattle. She could stand on her own, with reading and music. She did not get bored in my absence, and joined me to share the buffet lunches. She was a pretty sixteen-year-old girl, an age still chaste at the time. She befriended a son of Bott who was her age and some young participants, just enough not to feel too lonely and improve her English. At the end of summer 1967, Michelle enrolled in Orsay University to begin undergraduate studies. After some hesitation she chose mathematics, for which she was gifted and had a taste for. She was glad to leave school discipline, the narrow-mindedness of some teachers and to be able to work on her own timing. Of course, she also appreciated the mixture of girls and boys she had been denied from interacting with before.

Family House in Lubac

In 1965, Gustave, despite the financial problems created by the buying of our house in Antony, decided to buy or to build in Lubac, or its immediate vicinity, a family house where his children, and also his sisters, could come and have long stays with us. He finally decided on a large plot at the end of the village, at the foot of the mountain. The landscape is beautiful, Robert's hotel located only a few hundred meters away. The Le Roux village, its bakery and grocery stores (which don't exist anymore today) were about one kilometer away, an easy walking distance. The Saint Maurice village, at three or four kilometers away, and the small town Saint Firmin, at about ten, can easily be reached on bicycle, and of course, by car. The only inconvenience is that Lubac, as its name says (ubac means the slope in the shade; the opposite side receiving the sun is called adret) does not see the sun all winter, so the stay there is not pleasant at Christmas. There are no facilities for skiers. This deficiency did not bother Gustave, who had not skied in his youth and did not wish to start at fifty.

While Michelle and I spent July in America, Gustave, Daniel, Geneviève, Agnès and her son Philippe breathed mountain air in a rented house in Lubac. Gustave oversaw the building of the great house of his dreams, near the hotel of our friend Robert. When it was completed, we went there every summer, with Daniel and Geneviève of course. The children of our blended family and their children came regularly, as well as Gustave's sisters, Marthe and Marie-Thérèse, with her husband and their three children.

We also received some friends in Lubac, former students of Gustave. A few years later we took with us the daughter of the relativist William B. Bonnor, whom we called Bill, and who was Professor at the Queen Elizabeth College of London University. I had made Bill's acquaintance at a seminar directed by Mme Tonnelat. I had liked the precise results of his talk and his very clear presentation. We became and remained friends because we appreciated each other. We did not do work in collaboration, but Bill has always shown interest in my results, defending them, sometimes, better than myself. He told me that listening in London to a conference of a visiting Irishman about the Einstein equations, he had asked, "Is not part of that work from Yvonne Choquet?" The Irishman, who had not quoted my name once, had answered, "Yes, almost all." Bonnor came several times to Paris and I visited him in London, giving some lectures at the Queen Elizabeth College. His wife was a courageous and nice woman. Handicapped by a deforming arthritis since childhood, she was, however, always cheerful. She told me one day that she suffered constantly, but she had understood in her youth that it interested nobody, but drove everybody away, therefore one should not complain. It is good advice. Her husband did naturally for her and their two children what she could not do herself. Our families became friends. When our children, my son Daniel and Helen, the daughter of Bill, became adolescents, we exchanged them a summer for their linguistic progress. Daniel returned, happy, from London thanks to the kindness of Mrs. Bonnor. I am afraid Helen Bonnor did not keep such a good memory of her stay with us in Lubac. It was difficult to get her out of her room where she read English novels she had brought. We have not renewed the exchange of children, but the relations between the parents have remained as good.

Winter Holidays

As I said before, Lubac is unwelcoming in winter. For the Christmas holidays, Gustave, Daniel, Geneviève and I did, for a few years, travel to countries with more pleasant climates. After the Balearic Islands we went to Malaga, a little disappointed upon our arrival that the fishing hut on the beach we had rented through a travel agency was in fact an apartment on the fifth floor of a building. We wanted to comfort ourselves with an excursion in Grenada but I don't keep a pleasant memory of it. The visit of the Alhambra under the snow with an icy wind, that Gustave refused to shorten, got me a good cold. A better memory is the trip we took another year in Tunisia, where we visited the impressive ruins of Carthage, then went down along the coast. Kairouan was where we bought — and had sent to Antony — the rug which makes the pleasant comfort of our living room. The beautiful city of Sousse was where we bought, for Geneviève, a small straw camel covered in white skin, which she called Sousse. I also liked, a few years later, to see Morocco again. Then going south into the desert after visiting Marrakesh and Essaouira, I bought there a nice mother-of-pearl inlaid tray. Our last Christmas family trip was in the Canaries islands, where Daniel did not come and the weather was rather gloomy.

Research, Non Strictly Hyperbolic Systems

Leray published, in 1964, a long and difficult work, done in collaboration with the Japanese Ohya, and titled "Systèmes Hyperboliques Non Stricts". This article proves a remarkable theorem relative to systems of partial differential equations which do not satisfy the hyperbolicity criteria previously defined by Leray or by Friedrichs, but for which the Cauchy problem is nevertheless well-posed, with a solution obeying a propagation law which may imply causality for relativistic equations. However, the properties of the solutions of non strictly hyperbolic systems (called now "Leray-Ohya hyperbolic") are very different from those of the solutions of hyperbolic systems previously defined. They have an infinite number of derivatives which must verify inequalities, weaker than those verified by analytic functions. They cannot be discontinuous, but they may vanish outside of a domain without being zero everywhere. The Leray-Ohya hyperbolic equations at once seemed to me adapted to the solution of relativistic dissipative phenomena, where no shock occurs. The Leray-Ohya article of 1964 concerned only linear equations. It seemed to me that their results should extend to quasi-linear systems, but this extension is a big job, I was not ready to take it on, and Leray and Ohya were doing it themselves. The detailed writing of the article containing the proof of their result was published three years later. I said before how Lichnerowicz had used this result, before the publication of the article, to prove the non strict hyperbolicity of the equations of relativistic fluids with infinite conductivity. In fact, I thought that these equations were likely strictly hyperbolic; since these fluids are non-dissipative, shocks take place in them. They don't satisfy the Leray hyperbolicity condition, but this condition is not necessary, merely, sufficient. I suggested, to a student who came asking me for a subject for a thesis, to try to prove that these equations can be written as a symmetric system, Friedrichs hyperbolic. The subject was too difficult for the student, but Friedrichs' himself proved the theorem a little later, in Special Relativity. Still later, general methods of symmetrization have been introduced by Guy Boillat and Tommaso Ruggeri. A clear presentation is given in the very complete book by Marcello Anile "Relativistic Hydrodynamics and Magneto-hydrodynamics" (Cambridge University Press, 1987).

In 1965, one knew that the Leray-Ohya hyperbolicity was useless for fluids with infinite conductivity. I thought that, on the other hand, it could be relevant for many other relativistic matter where dissipative phenomena which should not destroy causality appear, fundamental in the Einsteinian conception of reality. It was fairly easy to prove that perfect fluids with finite, non-zero electric conductivity obey a Leray-Ohya hyperbolic system, as could be foreseen from the absence of discontinuities in flows of such fluids: for example mercury which, I was told by physicists, behaves like jam under the impact of a pebble. Viscous fluids are a more delicate problem. If one generalizes brutally the classical Navier-Stokes equations for viscous fluids, one finds an infinite propagation speed for perturbations, incompatible with relativistic principles. Many models more or less complicated have been proposed, all unsatisfactory. I proposed mine — it has the merit of being comparatively simple and of Leray-Ohya type, which is physically satisfactory. It did not have great success. It is true that reality is more complicated than all models, as soon as you go further than the first approximation. Moreover, today, numerical calculus on computers often replace the equations. Others besides me, such as Guy Pichon and Lise Lamoureux-Brousse in particular, have shown the Leray-Ohya hyperbolicity of systems satisfied by various types of matter in general relativity.

12

AFTER THE REFORM 1968–1979

The 1968 Revolution

The school year 1967–1968 began without problem for the Choquet–Bruhat family and apparently, also for the university and French society. I had been led to accept, for the second semester, a course of undergraduate mathematics attended by so many students that it had been split in two. Students whose names began with A to L followed my course, the others went to that of a colleague, an excellent mathematician not interested in physics. The lecture halls were calm and studious, however, agitation brewed and blew up with the warm days. Others have already written a lot about the May 1968 events which, starting with protests in university, spread to a general challenge to a too-rigid and compartmentalized society. I will only speak of my own experience. Living in Antony with my family, I did not witness any of the violence that occurred in the evenings at the Quartier Latin. The rebellious nature of my daughter Michelle, a student in Orsay, could have encouraged her to participate in this agitation; in fact, she took advantage of the suspension of classes to escape and spend some time in Les Landes with a companion of her choice.

Indeed, classes had been cut, replaced by meetings of discussions between students and professors. I was quite in favor of such discussions, which took place, with regard to me, in a relaxed and friendly atmosphere. Students told me only kind things; they were apparently satisfied with my teaching. They blamed the colleague who shared it with me for too-perfect a course which, consequently, was not challenging enough. My colleague promised to correct this fault.

Professors and students essentially agreed about general problems concerning curriculum and exams. I pointed out to the left wing of the student's assembly that they were the employers, and us, professors, the workers in their service. The trouble was that, from one meeting to the other, we had again the same discussion, starting from zero. It seems that it must be so in the permanent revolution, the ideal of a new society. I quickly got tired of this repetition and stopped going to the meetings. No penalty was given to me for my absence.

The universities' reform was decided among the authorities. I will speak only of what touched me closely — the split of the sciences faculty of Paris in two universities, Paris 6 and Paris 7, and the resulting explosion of the mathematics department of which I was a member. The split was essentially politically and personally motivated. My brother, François Bruhat, took in it an active part, mostly because of his wish to see a multidisciplinary university created. He had always had varied interests, in sciences but also in letters, philosophy and especially history. The Paris 7 University, known as Denis Diderot today, was designed, under the patronage of its President Alliot, to be such a university. It was intended also to be more democratic. In the mathematics department, the choice of professors between Paris 6 and Paris 7 was mainly determined by their political opinions. Gustave was a leftist; my brother François wrote to him, asking him to join

Paris 7. But Gustave had the spirit of contradiction and was jealous of his independence; he preferred to remain in Paris 6 where he obtained a research team independent from other members of the mathematics department. It was difficult for me to choose. Fortunately, Paul Germain, an apostle of "the true mechanics", took advantage of ongoing reforms to obtain the creation of a unit of mechanics independent from mathematics. This unit had no political color. It brought together a number of members of the old mathematics department interested in their applications. I asked for my admission to Germain. He granted it to me out of friendship, though I was not "a true mechanician". He also admitted my friend Avez. Later, describing what was called "UER (Unite Enseignement Recherche) de mécanique", he wrote, "This unit contains 18 mechanicians and 2 marginals."

The unrest of the students and later of the society in 1968 brought great changes in universities. Mandarins had to be removed, therefore Chairs should be suppressed. This done, courses were distributed anew among professors. From then on I shared my teaching, undergraduate classes, between the mathematics UER where I chose to treat generalized functions called in France Schwartz distributions, and the mechanics UER, where I was in charge of analytical mechanics. The Minister of Education of the time, Edgar Faure, among other regulations, decided that professors would no longer have power over the choice of their assistants, who were now appointed directly and for life, by *ad hoc* committees. I did not have to complain about the assistants who were attributed to me, but I stopped directing their research and feeling responsible for it. An unpleasant aspect of the reform was the multiplication of meetings, where time was lost in idle discussions between teachers and delegate students. I remember a meeting where we discussed something, I don't remember what, relative to exams. The professors agreed on a decision made on common sense. After a fairly long discussion, the delegate student told us, "I agree with you, but I shall vote against it, because it is the wish of those who have chosen me to represent them." To some other discussions, one waited in vain for a student representative. One day, when I pointed out to Lichnerowicz the disadvantages of the Edgar Faure reform which weakened the prestige and responsibility of professors, Lichné answered, "Edgar Faure is clever, he puts the mess in the university rather than in the street." It is true that, being in the Collège de France, he was not directly concerned.

Installation at Place Jussieu

The reform of universities coincided with the move of Paris 6 and Paris 7 to the newly constructed buildings at place Jussieu, on the site of the former wine market.

These buildings have been much criticized. They are spaced towers on a large esplanade swept by wind, icy in winter, little conducive to friendly discussions. The lecture halls are away, a fact which gives the advantage, according to some teachers, of not being bothered by students. Personally, I am always inclined to see the bright side of matters. Moreover I knew well the president Zamansky, a Jew who was grateful to my father for having helped him to survive during the war, and whom I saw spend a lot of time and energy for the accomplishment of these buildings. His merit has been recognized later by the naming of "Zamansky tower" the great tower dominating them, bitterly criticized years before.

As for me, I was glad when, arriving in Jussieu, I had a large office on the third floor of the tower 66 and an attached small room with a wash basin. This comfort had been planned for the professors by the efficient director of the UER of mechanics,

Vichnievski. My friend Avez occupied the next office. I also obtained, in front of these offices, a room which was used as a secretary office and library for the relativistic mechanics team, associated with the CNRS, which I directed since regulations had obliged Lichnerowicz to give up this position. Under his direction, the team had neither locale, nor credits nor secretary. It is true that the former administration was less paper-loving, and publishers of books or journals accepted and even preferred manuscripts that "protes" composed with "leads".

Unfortunately the "true mechanicians" refused to have the other offices on this third floor be given to relativists, hypocritically saying that one should not establish a segregation but favor interactions between researchers of different specialties. This argument seemed valid to me and I accepted our dispersion. I regretted it when I found that the "true mechanicians" had reserved for themselves the entirety of the fourth floor, the members of our team being dispersed with various others in the lower floors. This situation was unpleasant but not catastrophic. Good relations reigned in our team, partly thanks to Lichnerowicz who knew how to be respected as well as voice his wishes, while remaining friendly and warm.

We had the good fortune to recruit two secretaries who gave us invaluable help. One was Mme Goudmand who was in charge of administrative problems as well as of welcoming and housing our foreign visitors. She took her job to heart. I saw her regularly at the university, but she went to Lichnerowicz's apartment to perform secretarial work for him. Mme Goudmand has played an essential role in the cohesion of our team, as she took a personal interest in all the members. Our second secretary, Mlle Delègue, Mme Serot Almeiras, had training as a librarian. At first she showed some reluctance to type our mathematical works, but she resigned herself through necessity. However, she started to like it when the emergence of new tools, first ball typewriters and later computers, made the job more interesting for non-mathematicians. She learned to use computers, the Mathor program, then called Scientific Workplace. She was helped by a member of our team, gifted for these new techniques, Jean Paul Duruisseau. Jean Paul became a specialist of numerical and formal calculus on computer. It brought to him a well-deserved promotion as Professor. I myself have benefited from the advice of Duruisseau and of Mme Serot to start learning, before retiring, how to use these new tools.

University Life and CNRS

After the 1968 reforms, I was a member of the mechanics department. I taught analytical mechanics to undergraduates. My colleagues never allowed me to teach relativity. They also refused a course on the foundations of quantum mechanics among options proposed to graduate students. I would have liked to teach this option, teaching being the best way to learn. I was not given this possibility. The mathematics UER was, on the other hand, favorable to the creation of a "distributions theory" option which I taught for several years. This course led me to write a book published by Masson: "Distributions, Théorie et Problèmes", long out of print. Part of its content, translated into English, appears in the book written in collaboration with Cécile DeWitt "Analysis, Manifolds and Physics". The original print of my book on distributions had received, though I do not remember who it came from, nor which journal it had been printed in, a review reproaching my book for its clarity; this comment reminded me of the shadocks, "Why make it simple when one can make it complicated?"

I do not have much to add regarding my teaching job which continued until 1990 in the university now called Pierre et Marie Curie, except that I have seen the students' standards lower from year to year, in such a dismaying way that I retired when I was 66 (after 47 years of service because the years in the ENS count in the activity of teachers) without asking for the "over-number", which would have allowed me to teach for three more years.

I was and always remained a marginal: at the university I belonged to a mechanics unit, at the consulting committee of universities I depended, from the mathematics section, in the CNRS I belonged to a team associated with the theoretical physics section. I was thus led to sit, several times, by election or nomination, in meetings where one or the other of these organizations discussed nominations or promotions. These meetings gave me the opportunity to better know some colleagues, and to befriend them. I particularly remember, with pleasure, the CNRS commissions which were presided by the future Research Minister, Hubert Curien very efficiently. I am grateful to him for the friendship he showed me then, and later when we were both in the Académie.

I remember, with amusement, some spats with colleagues. One of them was with Dieudonné, a famous mathematician and member of the Bourbaki group who had an authority increased by a powerful body. I was supporting the inscription on the list regarding the ability to teach in universities, of a young mathematician, Guy Pichon, who had done interesting work on mathematical problems for the motion of fluids in General Relativity. Dieudonné opposed it, giving as an irrefutable argument from his point of view, "I don't know this work." I replied sharply, "You do not know everything." Dieudonné was an honest man — he ceased to oppose the inscription of

Pichon who made a fine career as Professor and researcher. Another altercation was at the CNRS committee. Lurçat, a physicist, though a theoretician using a lot of mathematics, violently attacked researchers he judged too mathematical. These systematic attacks upset me and I made unkindly remarks to him. At the end of the session I was in the hallway near Lurçat and the President of this meeting who said, "Now, shake hands." Lurçat answered, "With pleasure, I love getting yelled at." I replied, "Not me" and did not hold out my hand.

Private Life

I was now comfortably settled at the university with full-time secretaries to help with administrative or scientific works.

My family life followed its course. Our competent Agnès, pregnant again, had left us to marry the father of her future child, who was also the father of the first one. Once married, she stopped childbearing. We learned it when we met her by chance; she did not try to see us after leaving us. She had been marked by her past in Martinique and, in spite of the consideration with which we always treated her, could not quit the idea that we were enemies, like all whites. Her son Philippe had had mumps at the same time as our children, and I had shown him, like them, to Dr. Lepintre. There is no medication for this illness, but Agnès said (and believed), "He does not prescribe medication to Philippe because he is black."

After Agnès left, we made a few inconclusive tries to replace her: an "au pair" English girl, the offspring of a clergyman who was not surprised when we sent back to him his daughter, who was unable to take care of children or to wash dishes, then a young French, with only an older sister as family who turned her against her bosses, who, according to the sister, were necessarily exploiters. However, she got along well with Michelle. It was she who left us suddenly, following some grievance, perhaps because I had refused to provide to her for breakfast the porto wine prescribed by her dieting? She wrote to me a few months later, asking me to take her back. She was a girl to pity, but I answered the truth: I had already hired, in the meantime, a young Portuguese who gave me entire satisfaction. Palmira, only fifteen years old, had come to see me, presented by her mother who had boasted of her qualities. She had already been working for two years, taking care of the children, doing cleaning and cooking, but her bosses did not pay her. The mother looked nice, Palmira was smiling and robust, so I hired her. She fulfilled so well her mother's promises that she remained full-time with us until her marriage six years later. Palmira spent the weekends with her family but was in charge of cleaning and meals on week-days, and of bringing the children to school when I could not do it myself. Palmira and my daughter Geneviève, who was then three years old, liked each other. I often found them together in the kitchen. Her relations with Daniel were without problem. I think Palmira has good memories of these years of youth, laborious but carefree, and of her stays with us in the Alps where we made her share our love of nature. After the marriage of Palmira, we hired, upon her recommendation, again full time, another Portuguese, Fatima. Fatima was honest, serious and was not reluctant to work, but she did not have the pleasant character of our previous help. Palmira had had a happy childhood, surrounded by a united family, with parents who loved her. Fatima did not speak much of her childhood. She only told me that her life was difficult, being forced to work hard in the fields. The children did not form affective bonds with Fatima, but she did for them all the necessary material services. She was our last full-time employee. When she left us to marry, Gustave said that our children were now big enough to remain alone for a few hours. I agreed. Afterwards we were satisfied with a domestic help a few hours a week, first a not-so-young Portuguese, Rosa, recommended by Palmira. Rosa, always even-tempered, was flawlessly performing housework and was more than perfectly honest: having stopped working at one of her employers who had died, she was told she was entitled to a fire indemnity. She began by refusing, saying, "But he did not fire me, he died." Finally, I believe the heirs proposed a compromise that she accepted. Rosa worked for us until her retirement about ten years later. She then went back to Portugal and never missed sending me a nice card for Christmas. I asked Palmira again for her help in finding a housekeeper to replace Rosa. I was pleasantly surprised when she offered herself for the task. Her children were grown up, her time schedule left her availability, her home was only a few kilometers away from ours and she drove her car. Palmira thus became again a precious help for us, and remained, as well as her husband Joachim who took care of the garden, a faithful friend for the whole family.

Scientific Research, Friends and Travel

General Relativity had become an important part of physics. Without taking account of the meetings every three years of the society "General relativity and Gravitation" — officially created in 1971 at the congress in Copenhagen — international relations had multiplied. The following pages will be devoted mainly to my research work, an outline of the topics that have captured my interest, the personalities that our common interests made me meet and befriend and the travels I have done, often thanks to them, around the world.

Progressive Waves, First Order Systems

I said before that during my July 1964 stay in Seattle, I succeeded in extending results of a fundamental article by Leray, Garding and Kotaké on solutions with analytic Cauchy data of linear equations, to non-linear equations. It is always as much of a pleasure to find a new mathematical result, as to discover an unknown landscape. However a problem posed by physics has a greater attraction for someone who would like to understand the world we live in. An analyticity hypothesis is too restrictive for the study of a physical system, especially in the relativistic case where it disagrees with causality. On the other hand, a linearity assumption is unsatisfactory, because it implies that effects add without interfering, which is seldom the case in reality. The General Relativity equations are non-linear; it is already so for the Euler equations which rule classical fluids motion, and also of course relativistic fluids. Though the GKL (Garding, Kotake, Leray) article considers only linear equations, I had already noticed that the asymptotic expansions it defines can be used for non-linear equations. I applied the method introduced by Leray for the study of high frequency waves that gave solutions of systems of linear partial differential equations, to the case appearing often in physics of first order quasi-linear systems. I constructed approximate solutions of such systems involving high frequency waves. I showed that the non-linearity implies, in general, a stiffening of signals carried by the waves, then a shock: the built solutions become infinite after a finite time. I proved that it happens for relativistic perfect fluids. There are exceptional cases. The famous mathematician Peter Lax, Professor at the Courant Institute in New York, had met such cases in his study of shocks for solutions of quasi- linear partial differential equations in two variables. The French Guy Boillat generalized, to higher dimensions, the study of shocks and exceptional cases. Guy Boillat is an excellent unorthodox mathematical physicist. Originally a student of Louis De Broglie, he was more interested in rigorous mathematical results about problems linked to physics than in the speculations of the theoretical physicists of the sixties coming from his master's school. He is an original and independent spirit. He pursued, with success, research on non-linear shock waves propagation and also on the hyperbolicity of partial differential equations systems, often in collaboration with Tommaso Ruggeri, a professor in Bologna. Boillat was a professor in the mathematics department of the Clermont-Ferrand University, until his retirement when he returned with pleasure to his ancestors' house in his native city of Besançon, to which he remains very attached. His little, friendly letter upon the occasion of my ninetieth birthday was a pleasure for me.

As for me, I studied, beyond the case of a simple characteristic, the more delicate case where the system has a multiple characteristic. I used for this study some computations made by Jean Vaillant, who, like me was a student of Lichnerowicz that became a student of Leray. The results are then different and varied. My works on these general asymptotic waves have appeared in a fairly long article in 1969 in the "Journal de Mathématiques Pures et Appliquées". I lectured on them the same year in Bucharest at the congress "mathématiciens d'expression latine". I was flattered by the interest shown in my conference especially by Dieudonné who had the reputation of being hard to please.

Back to High Frequency Gravitational Waves

The propagation of gravitational waves is an important problem for our knowledge of the cosmos. The study of such high frequency waves was taken over in 1967 by an American researcher, Richard Isaacson, a student of Charles Misner, who apparently ignored my 1964 work. In his construction, Isaacson did not use the general developments of Leray, but the original WKB form. He took the oscillatory perturbation of the background metric such that its effects are greater than those of the background. He then found what he called a "back reaction". He computed the perturbation by an artificial splitting of the Einstein equations in what he called "wave equations" and "sources", splitting, in fact, not coherent with the magnitude of the considered terms. Besides, he could only conclude in terms of averages that were not clearly defined. The computation of the Einstein equations for such a metric is rather involved, but the results I obtained for the asymptotic validity of solutions of high frequency are rigorous and fairly simple.

I gave, in 1968, some lectures on gravitational waves at the University of Rome, invited by Professor Carlo Cattaneo. I had met him thanks to Lichnerowicz, who was a great friend of his. Both of them were admirers of the great Italian geometer, Enrico Bompiani, called "haricot" by Michelle in 1954. Cattaneo was quite a valuable mathematical physicist, the author in particular of a relativistic heat equation which remains the best approximation for this difficult problem. He was also a man of great human value, like his wife Ida. She had been studying mathematics, and abandoned it upon her marriage which was followed by several children. Ida had the misfortune to catch a viral hepatitis while pregnant with her last daughter, who consequently was, according to her mother, "born stiff". For four years, Ida devoted herself day and night to the care of her child who, Ida told me, was beginning to get better when she died from pneumonia. Ida fell then into depression. She got over it, on her husband's advice, by returning to work on mathematics.

Ida was interested in geometry. It was Lichnerowicz who guided her to do a thesis that gave her back a taste for life. Ida obtained a position as professor in Lecce University. She invited me in 1972, to spend a month in that beautiful city full of charm. There I gave a course on degree theory. I keep from it very good memories, and also of our descent to Tarento through Puglia. Later, Ida became professor in the new university of Rome, Torre Vergata. She invited me there and I was happy to see her again, devoted to her work and her children, always active and courageous, despite the untimely death (from a stomach cancer diagnosed too late) of her husband with whom she was very close.

After my lectures in Rome in 1968, it was Easter break in France. Gustave spent them in his house in Lubac with his children and his sisters. He encouraged me to take a few days of rest alone. On the advice of my friend Pham Mau Quan, I went to spend these few days in a pleasant hotel he knew in Ostia. I took this opportunity to finalize my results on high frequency gravitational waves. I proved the existence of such waves and their propagation with the speed of light for background metrics satisfying some conditions linked with the energy of the perturbation they create.

An interesting property of these gravitational waves which I discovered is that in their case, in spite of the non-linearity of Einstein equations, there is no stiffening of signals. Another aspect of this remarkable feature of Einstein gravitation, which I have highlighted later, is used in the proof of existence of global in time solutions of the vacuum Einstein equations with initial data close to the Minkowskian one by Lindblad and Rodnianski, different from the one given before by Christodoulou and Klainerman.

I wrote up my results of 1968 for a paper presented to the Academia dei Lincei by Pr Cattaneo. The rigor of demonstrations leading to clear results, original and remarkable, made of this work a pleasure: in spite of it, these few days in Ostia put me back in shape after a tiring quarter. After my return to France, I completed, on these gravitational waves, a detailed article published in "Communications in Mathematical Physics". Soon afterwards, I had the occasion to meet Richard Isaacson. I found a nice young man who held me no grudge for having taken anew and corrected his work. He moved later to the administration and became responsible for the "General Relativity" section in the NSF (National Science Foundation). I am grateful to him for the CNRS-NSF agreement which favored my collaboration with the remarkable relativist Vincent Moncrief.

I extended my results on high frequency gravitational waves to the coupling of these waves to such electromagnetic waves, showing that they generate each other. Lichnerowicz proved, later, an analogous property for the electromagnetic and gravitational shock waves that Taub and I had defined.

Detection of Gravitational Waves

I explained my work on gravitational waves at Haïfa in Israel in 1969, at a meeting organized to honor the sixtieth anniversary of Professor Rosen. After my talk, Joseph Weber congratulated me warmly for my results. In fact, he had heard them a few months before. So I apologized that my lecture was not new. He answered, "New for me anyway", a blow to my vanity! I had established, some years before, friendly relations with Joseph (called Joe) Weber, invited by Lichnerowicz to the Collège de France to speak about his experiments for the detection of gravitational waves. He was a friendly man without ostentatious pride, though he was passionate about his work and apparently persuaded of the validity of his observations. The astronomers don't believe it anymore, all (almost) believe that gravitation propagates with the speed of light and there exist phenomena which can be called gravitational waves, but the weakness of these waves due to the smallness of the gravitational constant in usual units makes unlikely their detection by an apparatus of the type constructed by Weber. The relativists are, however, grateful to Weber to have been the first to try this detection. The following questions remain open: Did Weber have a mysterious gift owing to which these waves were revealed by his apparatus? Did he himself believe in the positive results he announced? A fact which may lead to doubting it is that after having announced a twenty-four-hour periodicity for his observations, as for the earth rotation, he brought it to twelve after it was pointed out to him that, in opposition to electro-magnetic waves, gravitational ones go across matter, therefore his laboratory received the brunt of them twice a day. However, in 1969, Weber was respected by all relativists. We all sympathized with him when he learned, through a telegram, of the death of his wife from a heart attack. The Webers were a nice couple who were very close. Life follows its course; Joe remarried some years later with a young talented astronomer, Virginia Trimble, who was a faithful believer in her husband's work. Virginia was nevertheless a woman of independent character. One day, when I presented her to a colleague as "Mrs. Weber", she corrected me with energy, saying, "I am Virginia Trimble." She was a staunch feminist. I am grateful for her intervention at a meeting of the GRG Society in 1982, where the names of future possible presidents for the three years to come were discussed, all men. Virginia said, "Before discussing this subject we must thank a woman, Yvonne Choquet-Bruhat, for the efficiency with which she served as President during the last three years."

Joseph Weber and Virginia Trimble have been pioneers in pursuing the detection of gravitational waves. Relativists are grateful for their faith in the possibility of this detection which has been a motor for research. The governments of the United States on one hand, of Italy and France on the other hand, have invested considerable sums to build devices called LIGO and VIRGO. These are nothing like the Weber cylinder but have, after fifty years of effort, a sensitivity sufficient to record, on Earth, the oscillations of a gravitational wave due to the shock of two black holes, of respectively 29 and 36 times the solar mass, located at more than one billion light years away. The shape of the received signal confirms the Einstein theory of gravitation (General Relativity), because it coincides with the one predicted by Thibault Damour and his team after very complex algebraic and numerical calculations made during many years. The building of several gravitational wave detectors is planned on Earth (India, China,...). A detector in space, LISA, is also planned. It will reach a sensitivity unmatched on Earth.

Tourism in Israel

At the 1969 congress in Israel, I took along my daughter Michelle, then eighteen years old. Well-meaning souls told me that, given the political situation and the attacks, it was not prudent. In fact, at the time the country seemed quiet to me. We arrived in Tel Aviv, a big city without charm, and went from there to Haïfa, a beautiful city full of flowers where the congress took place. From the university, which was situated on a hill, one has a splendid view of the sea. Michelle befriended some Israelis and accepted their invitation to stay a few days after my departure. I had to resign myself, although not without concerns which proved unjustified. As for me, I stopped for two days in

Jerusalem, where I made unforgettable visits to Christian and Muslim holy places. There was a striking contrast between the Jewish Jerusalem, an occidental city, and the Arab Jerusalem, full of picturesque visits of hills covered with olive trees, whose charm seems to have forgotten the tragedy that took place there. I stayed in a small hotel with a sunny and flowered courtyard. I took a taxi to reach the airport, and held a nice conversation with the driver. I did not note his religion.

Cambridge

In July 1976, a program was organized in Cambridge (England) on general relativity. I was invited as well as most of the specialists in the mathematical problems linked with General Relativity at that time. Gustave was leaving for holidays in Lubac and refused that I take Geneviève to England where, he said, it rained all the time. On the other hand, he could not oppose that Daniel accompany me and so, perfect his English.

The other participants were accommodated in university buildings, but the law of the time forbade women this privilege. The organizers found, for me, a nice house in the neighboring country. I rented bicycles for Daniel and me; it was perfect. Unlike the prognosis of Gustave, the sun shone during the whole of our stay. Daniel went on bike rides and joined me in the refectory for lunch; he thus had occasion to practice his English. In this program participated Hawking, with whom we could still converse, and his student Gibbons, who had since become a master and took care of him at meals. Taub was there and we took that opportunity to strengthen our friendship by collaborating on an article. Abe Taub also had been interested in high frequency gravitational waves, but had chosen to use a Hamiltonian method that was dear to him. We decided to treat together the open problem of gravitational waves coupled with scalar waves. Taub was kind enough to endorse my method for that work. It is true that it gives more accurate results. Our collaborative article was published in 1977 in the newly created "General Relativity and Gravitation" journal.

I returned several times to the Newton Institute in Cambridge where several very valuable researchers are working on General Relativity. Among them are Malcolm Mac Callum, who directed for a long time the administration of the GRG Society, and Stephen Hawking, known worldwide for his work and his courage in overcoming his terrible dicease. In 2005 I spent two fruitful and pleasant months there on the occasion of a program about mathematical problems of General Relativity. Like other participants, including my friends Jim Isenberg and Sergiu Klainerman, I was housed in a small condominium apartment near the university.

Cargèse

I had as a colleague at Paris VI, Maurice Lévy, a theoretical physicist, who had freed himself from the scientific domination of an aging Louis de Broglie. With the support of Yves Rocard, he founded in the École Normale Supérieure a theoretical physics group of valuable young scientists, including Louis Michel who became the first full-time professor at the IHES (Institut des Hautes Etudes Scientifiques). This institute is dedicated to advanced research in mathematics and theoretical physics. I will come back later to this marvelous institute where I have had an office since 2003. Maurice Levy's group has received distinguished visitors, including Gellmann, the discoverer of quarks. I once happened to travel sitting next to this future Nobel Prize recipient in a train, at a time where the existence of quarks was still speculative. Gellmann was pleasant and talkative, absolutely sure of the truth of his discovery. In addition to his teaching and his research, Maurice Lévy has worked hard to promote, in France, the development of theoretical physics that was in full renewal in the world, in particular in the United States. He played a fundamental role in the creation, in La Villette near Paris, of a scientific museum and activities center, "Centre International des Sciences et de l'Industrie", which is very popular now. Before that he brought an essential contribution to the creation, funded by NATO, of an International Center of Theoretical Physics in Cargèse, a Corsican village. In 1978, Maurice Lévy, helped by Stanley Deser organized there a session on gravitation. I was invited to teach a course entitled "Introduction to General Relativity". I accepted with pleasure. Corsica is a magnificent country, where sea and mountains compete in beauty. The Theoretical Physics Institute is located on the beach front in a vast area covered with wild vegetation. It enjoys a private beach. I requested to be accommodated in the Institute like the students. It was convenient and pleasant. My friend and future Nobel Prize recipient Georges Charpak, a very wise man, had bought a piece of this land before its acquisition by the State, and had a house built where he spent a few holidays. Georges invited Cécile DeWitt and I to a nice dinner in town, on the occasion of the passage of the future Nobel Prize holder Lederman, with whom he had worked in Chicago. Both the DeWitts were in Cargèse then, making my stay even more enjoyable. Bryce was delivering his first works on the quantization of the gravitational field whose developments have made his fame and his Washington Academy election. Cécile had no lecturer obligation but was an assiduous listener. I appreciated her presence. They had the visit of their oldest daughter Nicolette, a brilliant girl who later, out of Harvard, had an important position in the World Bank. All four of us left on an expedition to the spectacular inland. We slept under the stars and the sky was beautiful, but I keep mixed feelings because it being located at high altitude and me having only a very thin sleeping bag, I shivered all night.

I stayed another time at the Theoretical Physics Center in Cargèse in 2002. Piotr Chrisciel and Helmut Friedrich had organized there a summer school titled "The Einstein equations and the large scale behavior of Gravitational fields", with the subtitle "Fifty years of the Cauchy Problem". The lectures at that school have been published by Birkhauser. In the original French preface of this book, the editors, Helmut and Piotr, call my thesis of 1952 "travail fondateur" (in English "seminal" or "breakthrough", not "funding" which refers to money, used by a lecturer to the amusement of the audience), and they dedicate the book to me with some flattering sentences for my subsequent work. At the end of the stay, a small party was organized and Bob Geroch was commissioned to make a speech. I do not seek honors, but the kind words of Bob made me happy.

Kinetic Theory

After my work on relativistic fluids, it was natural that I got interested in kinetic theory in the framework of General Relativity. This interest was stimulated by astrophysicist colleagues who were, at that time (1970), working on new theories about the beginning of the universe and using cosmological models where matter is in the simplest possible state, that is, it obeys the kinetic theory where it is assumed to be composed of point-like particles that interact only by their shocks and collisions. Such matter is represented as a function of a point and a vector that, one can imagine, without being a mathematician, represented particle density and speed respectively. This function is called a distribution function. In the absence of shocks between particles, the distribution function is conserved by their motion, and satisfies a homogeneous first-order linear differential equation called the Liouville or Vlasov equation. If there are shocks, the evolution of the distribution function is modified by a non-linear operator called the collision operator. It then satisfies an integro-differential equation called the Boltzmann equation, named after the Austrian scientist born in 1866, who was the first to consider it in classical mechanics. Around the time I began to be interested in the Boltzmann equation in General Relativity, the study of a particular model had led R. K. Sachs and A. N. Wolfe to some results in agreement with the discovery of the primordial cosmic radiation. On the other hand, J. Ehlers had resumed and completed some Russian (Chernikov) and German (Bitcheler) works on the Boltzmann equation in General Relativity (the case of Special Relativity was treated by Lichnerowicz and Marrot in 1941). However, no general mathematical theorem existed about the system of Einstein equations with source generated by a distribution function, coupled with a Boltzmann, or even Vlasov equation, satisfied by this function.

I had appreciated, back in 1957, the DeWitt couple's qualities, especially Cécile's energy. I had found unjust her situation of moving surrogate. I had her invited for a month at the Poincaré Institute; it gave her the occasion to seriously resume her personal research on the Feynman integral, fundamental in quantum theories. She was grateful to me and she had me invited for a month in Chapel Hill by her husband Bryce, without teaching duties. I took advantage of it to look for the solution of the puzzling open problem in relativistic kinetic theory. I first proved a uniqueness theorem for the solutions of the Cauchy problem for the Einstein-Vlasov equations. Bryce suggested, to justify funds of his research contract, that I write this first result for a publication paid by the university in the Journal of Mathematical Physics. I did this writing during that stay in Chapel Hill, embellished by interesting talks with Cécile DeWitt.

The existence theorem for the Einstein-Liouville equations, which I proved the following year, was published in "Annales de l'Institut Fourier". I extended the results to equations coupled with an electromagnetic field for an article in the book dedicated to Professor Petrov upon his retirement. These articles use somewhat different techniques, linked with the considered problems, but both are precise and complete.

The Boltzmann equation offers a more difficult problem. I waited until Daniel Bancel solved the Boltzmann equation in General Relativity in his thesis to write with him in 1973 an article proving existence, uniqueness and stability for solutions of the coupled Einstein-Maxwell-Boltzmann equations.

Daniel Bancel, a former student of the Saint Cloud École Normale (now Lyon École Normale Supérieure), is an excellent mathematician like my other student Guy Pichon. After his thesis he became a professor in the University of Toulouse, as he wished to be near a domain he owns in Lot. Thanks to his works, he was easily preferred over other candidates. He joined in Toulouse another former member of our team of relativists, Albert Crumeyrolle, author of an interesting thesis on spinors which had been directed by Lichnerowicz. Crumeyrolle had passed the Agrégation while teaching in primary school, and had started a thesis while a professor in high school. Lichnerowicz had got for him an allocation from the CNRS so he could finish it quietly. I had discussions often with Crumeyrolle, and we got along well. I had esteem for him as a mathematician and as a human. I think he also liked me. Upon the occasion of a feast at a "journées relativistes" colloquium he had organized in Toulouse, he had a decoration of the "Chevaliers de l'Armagnac" given to Lichnerowicz who showed great pleasure in it. Crumeyrolle apologized for not being able to give it to me as well because women were not admitted. This lack of the decoration had not disappointed me, but the kindness of Crumeyrolle's apology made me happy.

Crumeyrolle had an unusual career and was, I think, a bit isolated at the university. On the contrary, Bancel, in addition to his teaching and direction of researchers, quickly took an active participation in university life. He began by reviving a mathematical journal that had been famous but was somewhat off: "Annales de la Faculté des Sciences de Toulouse". He obtained, alone or with his students, new and interesting results on mathematical problems touching General Relativity or other subjects. His organizational and entrepreneurial skills had him quickly elected university president. He told me then that he would take care of administration for a while, before returning to research. However, his success as University President made him fairly fast chosen as Rector, first in Rouen, then in Lyon, and finally in Versailles. Important situations, useful to society, also honorary positions, not without material benefits. I had seen my uncle take advantage of them in Poitiers, then in Strasbourg. Daniel Bancel also let himself be tempted by these functions. I regretted it for science, but I respected his choice.

Einsteinian Constraints

While Bancel was working on the Boltzmann equation, I resumed the study of constraints in General Relativity, looking for a more geometric formulation than the one I used in Louis

Witten's book of 1962. I was invited to give a lecture at the International Congress of Mathematicians in 1971, in Nice where Dieudonné was a professor; Leray was President of the Congress. My conference was to propose a new way, based on conformal splitting, to solve the constraints equations satisfied by the initial data in the Einstein gravitation theory. A detailed writing of my work has been published in "Communications in Mathematical Physics". I had noticed that there was no general proof of the existence of the conformal factor, especially in the case of a compact initial manifold, for instance a three-dimensional sphere, a model often considered in Cosmology. I had found a course of Leray's on degree theory in the Collège de France — called Leray-Schauder — interesting. By using this theory, I proved an existence theorem for that conformal factor on a compact manifold. I wrote up my proof and brought it to Leray for its presentation as a paper in the "Comptes Rendus de l'Académie des Sciences". Leray, a very conscientious Academician, read it carefully and found it interesting, but he told me that the introduced method could probably be used in a more general case. He suggested that we write together a Note treating the general case. It was, for me, a great honor and I accepted with pleasure. Leray wrote the paper, and sent it to me for approval. I had the satisfaction to make a correction, minimal but useful, that Leray willingly made. Once the general case was treated, I wrote up the application to the Einstein equations. Leray presented this new paper to the Compte Rendus, but kindly declined my proposal to cosign it.

I wrote this work in more detail for the Acta Istituto Matematico, on the occasion of a lecture at a colloquium in Rome organized in 1972 by Carlo Cattaneo. Later the results have been generalized to non-compact manifolds by my student,

Alice Vaillant Simon, and myself. Alice, later Chaljub Simon, became a professor in Orleans and continued to obtain interesting results on partial differential equations.

Cécile DeWitt, Analysis, Manifolds and Physics

Cécile had studied and appreciated my book "Géométrie Différentielle et Systèmes Extérieurs" published by Dunod, and she offered to translate it in English. I thought it a pity for a physicist of her caliber to do a simple translation work. On the other hand, I wished to expand the content of this book. So I proposed to her that we write a new book in collaboration. Its mathematical content would retake some basic chapters of my books on differential geometry and distributions theory but would contain others and include important applications to theoretical physics — then in full development — either in the main text, or as problems with solutions. At the request of Cécile we took, as an associate for this first book, a student of hers, Margaret Dillard-Bleick. It earned our book, "Analysis, Manifolds and Physics" published by Elsevier in 1977, the nickname "The Three Women's Book". Margaret, a precise mathematical mind, was a useful collaborator, taking charge of the first draft of a number of chapters on differential geometry. Cécile and I, without the help of Margaret occupied with other tasks, revised and completed our book for a new edition in 1982. Our book has since been repeatedly reissued. I think that its success comes from the clarity of the exposition that always starts from bases known after mathematical undergraduate studies to arrive at recent results useful for today's theoretical physics. Cécile was a very efficient collaborator, always eager to succeed and never shrinking from a difficult calculation. Her good temper makes working with her quite enjoyable.

My Italian Friends, "Waves and Stability in Continuous Media" Meetings

At the colloquium of Rome in 1972, I met two young Italians, both speaking excellent French, who became my faithful friends: Giorgio Ferrarese, a student of Cattaneo, and Antonio Greco, who was then working in the Mathematical Physics department of Messina, directed by Carini, a friend of my former colleague in Marseille Vincensini. Giorgio took me to visit some of the most beautiful neighborhoods of Rome. Antonio told me he had already listened to me, with interest, the previous year at the International Congress of Mathematicians in Nice. With the agreement of Carini, he invited me for a month in Messina. I thus made my first visit to this land blessed by gods. Antonio is not a Sicilian by origin, but comes from Calabrese nobility. He has, of it all, the qualities of humanity and intelligence. Thanks to Antonio, and other Italian friends he introduced me to, I made many constantly pleasant stays in this southern Italy, rich in historical memories and wonderful landscapes. For my first visit, in 1973, I took a plane to Reggio di Calabria, then a boat for Messina. I stayed in a nice hotel at the seaside. Antonio came to accompany me to the university, and we talked about a little of everything. His enthusiastic youth rejuvenated me. We became friends for life. During my second visit to Messina, in 1975, I met another student of Carini, also very nice, Andrea Donato, but I knew him less well because he died prematurely.

After a few years in Messina, Antonio became a professor in Palermo, the capital of Sicily, where is located its greatest university. His happy disposition allowed him to adapt as an original Sicilian and continue, with efficiency, to do research and teaching. As for me, I went several times to Palermo; it is a beautiful city on the seaside, enriched with varied monuments and parks in bloom. I took advantage of one of my stays to go admire the Greek temple of Segesta, isolated in nature and splendidly conserved. I was the only passenger to return by train from a nearby station. I think it does not stop there anymore. I spent a week-end in Agrigento; it impressed me less than Segesta. I met the wife of Antonio, Argene, a literary who was amiable and intelligent. My stays in Palermo were made up of happy moments.

Antonio and I had a common interest in mathematical physics, in particular the matter sources of General Relativity, so we could collaborate on some joint publications on relativistic fluids and their interaction with gravitational waves.

Thanks to Antonio and to Tommaso Ruggeri, whom I met in 1981, I became a regular participant to the congresses they organized every two or three years with Salvatore Rionero, a professor in Naples University, under the title "Waves and Stability in Continuous Media". The subject of General Relativity found its place in these meetings, partly thanks to me. As for me, I was happy to hear talks about scientific subjects of interest to me even if on non-relativistic mechanics, in beautiful places often of southern Italy.

John Archibald Wheeler

I was with J. A. Wheeler at the General Relativity Congress of 1971, in Copenhaguen. I was on friendly relations with John Archibald since our meetings, first at Chapel Hill in 1957, then at the Batelle Institute in 1967. During the Congress in Copenhaguen he offered me an invitation for a month at the Princeton University. I accepted with pleasure. I thus spent a month of 1973 in this nice small New Jersey town, this time at the university instead of the Institute for Advanced Study. Jimmy

York was charged to find me a home. It was in the building of graduate students, a pleasant two-room apartment reserved to visiting professors. A refectory served meals, simplifying my life. There, segregation reigned: tables of black students and tables of white students without mixing. One day, seeing a free seat at a table of black students I sat at it to express my anti-racism. I was greeted by malicious glances. One of the guests came to sit next to me. He told me that black Americans had a hostile prejudice against whites that he did not share because he came from Africa and his ancestors had not been slaves. He added he understood that the blunder I made siting at that table was due to my French nationality. I did not do it again that year. Fortunately things have changed since: the separation does not exist anymore in the university refectories of the United States.

In 1973, the Princeton University had an excellent physics department. It still has one, but its members have changed. When I stayed there, Wheeler directed an important group of relativists: colleagues, students or visitors. I met there young researchers that have since become masters in the United States or their country of origin. I discussed much with Jimmy York who, inspired (he told me so) by my articles of 1962 and 1971, improved the formulation of the constraints equations in General Relativity by the method called conformal. This method had been initiated by a student of Georges Darmois, Racine, and partially extended by his other student, Lichnerowicz. It is currently used now under the name "Lichnerowicz-York conformal method". I find it a bit unfair that my name will be forgotten, but it does not make me sick. I always had excellent relations with Jimmy. In 1973, he told me that Wheeler had invited me so I would reassure him about the value of his student's work, which I have rightly done.

Positivity of Gravitational "Mass"

I had met Jerrold Marsden, my junior by twenty years, who was a doctor from Princeton University born in Toronto. He was a mathematician interested in numerous subjects, theoretical or applied. In Princeton, he had followed the course of my husband Gustave Choquet and had written it out with the collaboration of other listeners. Between 1972 and 1978 after meeting a researcher on General Relativity, Arthur Fisher, he worked with him on this subject. Their work, though it brought few really new results, had much success among mathematicians and contributed to interest them in problems posed by General Relativity. Marsden came to spend a month with our team in 1976. I attacked with him the problem of positivity of the gravitational mass of asymptotically Euclidean space-times, defined by Brill and Deser, with a heuristic proof of its positivity. We succeeded in making this proof rigorous, but only for space-times near to Minkowski space-time. I showed Jerry the falseness of several trials of proof he showed me to suppress this restriction. Our work in collaboration was published in the Communications in Mathematical Physics in 1976. Some years later this positivity was proven without restriction to a neighborhood of flat space-time, first by Schoen and Yau, then by Edward Witten, though two unrelated methods, both different from the Deser heuristic method which we had, in vain, tried to make rigorous in the general case. This difference reassured me about our failure.

In 2006, with the friendly encouragements and useful comments of Thibault Damour, I used the Witten method to give a mathematically rigorous proof of the positivity of the gravitational mass in higher dimensions.

Abraham Taub and Berkeley

Upon the occasion of various encounters, a true friendship was born between me and the couple Abraham (called Abe) and Cecily (called Cece) Taub. Cece told me one day that I was one of the few people they both liked. Abe, twelve years my senior, was appreciated also by his almost-twins Lichnerowicz and Papapetrou, as indeed by the entire community of mathematician-physicists interested in General Relativity. Educated in mathematics and Classical Mechanics, he had a rigorous and precise mind. It led him to fundamental results on relativistic fluids and solutions of the Einstein equations. We owe to him the building of an Einsteinian space-time, called "Taub space-time" that exists only for a finite time, though it has no singularity. Its geometric extension, built by Newman, Unti, and Tamburino, has strange properties which earned it the name "Taub-nut"; it is used as a counter-example to various conjectures.

Taub wrote to me in 1973, suggesting that I visit Berkeley where he was Professor. I replied that I would come with pleasure, but that in fact I would like it to be a long visit, one term, that is three months, where my husband would be also invited, and we would take along our children. Abe quickly sent me a positive answer for 1974, for me and also for Gustave, whom he had no difficulty inviting through the mathematics department where a statistician appreciated his work. Cece found for us a nice house not far from the campus, left vacant by a colleague visiting abroad. He left us also the au pair he used who was a great help for us, in particular, in taking care of our children when we could not do it. We enlisted Daniel (12 years old) and Geneviève (8 years old) in US schools. We also enrolled them in French correspondence courses and we ensured they learnt their lessons and executed homework, Gustave for Geneviève and me for Daniel, tasks which proved easy. Geneviève liked her American school, which was near to home. Daniel had some problem with his, where he got to by bus. In the United States of 1974 the plague which now afflicts some schools in France, racketeering, existed already. Daniel was victim for a while but learned to defend himself.

In Berkeley, I gave a course on General Relativity. I don't keep a vivid memory of it. I remember that Vince Moncrief, though already a confirmed researcher, was an assiduous listener. I remember also that every time I passed by the door of his office (overlooking the corridor, windowless on the outside), I saw him focused on his work.

After the term in Berkeley, us four Choquets bought camping equipment, including two tents, and we went to visit California, the parents sleeping in one tent and the children in the other. My memories are mixed with those of my travel in the same region twenty years before.

I met the Taub couple in most of the congresses on general relativity I went to, always with interest and pleasure. I was at a congress in New York with them and the Papapetrous when my hand-bag containing my passport was stolen in a restaurant. I called the embassy to obtain a paper to show to the police when leaving for France. I came across an unpleasant employee who told me that to obtain such a document I should come to the embassy with someone who had known me from my birth (!) or at least a French person who knew me for a very long time. Mmes Papapetrou (Greek but living in France) and Taub (an energetic woman when necessary) offered to accompany me in this expedition (New York is a big city). When we arrived at the embassy we were received by a very kind man who made, for me, the required paper without any objection. My stolen passport (not my bag) was found a few months later by the American police and sent back to me.

I stayed again in Berkeley for a month in 1984, invited by Taub who had learned that this university wished to receive, for this time, scientists of the sex formerly called weak. The Taubs found for me a comfortable room and I was happy to be again in this pleasant campus, and to discuss again relativity with Abe and various subjects with Cece. Despite their age of eighty years, the Taubs had the kindness to come to the colloquium organized in 1992 at the Collège de France, when I retired from teaching. Thanks to a researcher, Eric Lehman, a student of Lichnerowicz hosted by Abe in Berkeley and whose geologist father held an important position, they could visit the original Lascaux cave and were happy of it.

Hambourg and Bruxelles

During the years 1969 to 1974, I had the opportunity to make various short trips abroad to give lectures, often to meet a scientist with whom I had intellectually and humanly sympathized with. Such a trip took me to Hamburg in 1970. I enjoyed the rustling activity of this great harbor. I was invited by Wolfwang Kundt, a kind young man as full of energy as his city. He had done his thesis in 1953 in Hamburg under the direction of Pascual Jordan. Jordan was a great physicist-mathematician. He is said to have discovered the Fermi-Dirac statistics before the two famous men whose name it bears, but a blunder of Bohr's retarded the publication of his work for several months. After the war, Jordan did not get the reputation deserved by his scientific discoveries, including, perhaps, the Nobel Prize. I learned recently that the cause is his participation in the Nazi party during the war, which had caused his colleagues, opponents or Jews, to flee or die. He had not taken an active part in the denunciations, but as the daughter of a man, Georges Bruhat, who had done his best to protect the possible victims of the Gestapo and had died because of it, I cannot pity Jordan for his lack of glory. A contribution of Jordan's to general relativity is the improvement of the work of two other physicists, Kaluza and Klein. These, as others and Einstein himself, were searching a mathematical formulation uniting gravitation and electromagnetism, the two force fields known at the time, the way Maxwell had found a representation where the electric and magnetic fields were the two aspects of the same field called electromagnetic. Kaluza in 1921, completed by Klein in 1926, had shown that by adding a fifth dimension to the space-time's four and considering solutions of the Einstein equations independent of this fifth dimension, one obtained the coupled Maxwell and Einstein equations in dimension four, under the hypothesis of the constancy of a certain coefficient. Jordan, in 1947, wrote the obtained equations without this last hypothesis and interpreted the new coefficient as what was previously the gravitational constant, now variable. The geometric formulation of Jordan's work and the physical interpretation of results were proposed to Yves Thiry as the subject for a thesis by Lichnerowicz, who then named the result "Jordan-Thiry theory". The unification in a geometric formulation of gravitation, electromagnetism and the new fields (weak and strong interactions) called Yang-Mills, discovered after Einstein's death, was done by Richard Kerner, Andrzej Trautman and myself, by considering even higher dimensional space-times. They are called "Kaluza-Klein theories" — too bad for Jordan. He was still a great scientist. In addition to Kundt, he has trained many students, among whom Jürgen Ehlers, Englebert Shücking and younger ones who became brilliant mathematical physicists. I kept contact with Jürgen and Englebert, but I did not see Kundt again after my visit in Hamburg. He got more interested in the analysis of astronomical observations than in mathematical results. Looking on the Internet, I see that he has retained his independence of mind. I appreciated his criticisms of some beliefs common in the current environment of astrophysicists, though they were without a true observational or mathematical basis.

In 1971, I spent a few days in Brussels, giving a lecture at a colloquium organized at the Free University by Jules Géhéniau, its former Dean. Géhéniau was a former student and collaborator of Louis de Broglie who had done some work on General Relativity. He was a respectable and friendly man, and Brussels is a beautiful city, especially its main square and its magnificent cathedral. I went there again in 1974 for another congress, organized this time by the mathematician Michel Cahen, a specialist of symplectic geometry, who was appreciated by Lichnerowicz and a friend of Moshé Flato.

I went again through Brussels several times, in particular, coming back from Mons where the physicist Philippe Spindel, a longtime friend of Richard Kerner, had me elected as honoris causa Professor of the Mons Hainaut University, concerned with parity. There were several women among the newly elected but there was also my colleague in the Academie, Yves Coppens, who made an interesting presentation on his prehistoric findings, and a speech full of humor at the reception ceremony of the newly elected — I was relieved not to have to do it! I was happy to get to personally know Spindel and Coppens better than I had done at the university or the Académie. I was happy to meet them again later. I see Philippe at the IHES where he works with Thibault Damour on quantum gravity now. In coming back from Mons, I stopped in Brussels, hoping to visit a temporary exhibit of Magritte, but I should have reserved my attendance weeks before. I had to satisfy myself with viewing the permanent collections, but it was good nonetheless.

Greece and Crete

I spoke earlier of my trip to Greece with Léonce in 1956. In 1971 I had the pleasure to see Athens again, invited to the Balkanic Mathematicians congress. I had befriended the organizer, Professor Sterghiou, when I met him I don't remember where. Sterghiou, already about sixty years old, was responsible for a technical institute in Athens. He was an important personality preoccupied with the development of mathematics in his country. I keep a good memory of this congress, especially of my encounter with mathematicians from central Europe, then under soviet rule, and their wives, who could seldom go out of their country. They did not complain about their fate. One of them told me that their life was better since inheritance being suppressed, it was useless to save money to help one's children. Sterghiou was originally from Crete. His dream was the creation of a university there. As a first step towards this realization, he organized, in the summer of 1972, a congress on the teaching of mathematics. My husband, Gustave Choquet, known for his activity in this field, was appropriately a participant. Out of his friendship for me, Sterghiou also chose me as lecturer, thus justifying the refund of my travel expenses. I think that my talk about teaching was as interesting as many others.

In principle, Gustave and I never left our children at the same time. However that summer was an exception. My sister-in-law, Marie-Thérèse came with her children to Lubac and took charge of Daniel and Geneviève for a week. Sterghiou had organized our trip. Gustave and I took, at his expense, a plane for Athens, then a comfortable boat that brought us, in a night, to Heraklion. A bus then led us to Rethymno, a charming little city not yet denatured by tourism, where the congress took place, in a hotel on the seaside. This stay in Crete was very pleasant: the local people received us very kindly and friendliness reigned among participants. Some had been at the previous congress in Athens, others were well-known mathematicians interested in teaching. Sterghiou organized a folk show and an excursion across the beautiful landscapes of this island in the Mediterranean Sea. When we went back to Athens, he asked his son, a man about thirty years old, to take us to visit the city, which he kindly did. The university dreamed of by Sterghiou now has existed for many years. It has excellent professors and international fame, but, perhaps for political motives, Sterghiou's name appears nowhere; I find it somewhat unfair. Nevertheless, I hope that Sterghiou's soul, if souls exist, is happy to see his wish fulfilled.

Warsaw

In 1973 and 1976, I went to Warsaw in congresses organized by Andrzej Trautman. The first was for the five-hundredth anniversary of the famous astronomer Copernicus who put the Sun at the center of the system, around which turn the planets, including the Earth. The second was entitled "Geometry and Mechanics", a title suiting both Lichnerowicz and myself. Andzrej Trautman, with a Polish father and a French-born mother, is a relativist of international fame and a very agreeable man. He has done part of his studies in France and speaks perfect French. Andrzej restarted for relativists the fruitful relations that reigned before the war between French and some remarkable Polish mathematicians. The names of Banach, Kuratowski, Schauder, among others, remain famous. My husband, Gustave, finished the redaction of his thesis at the end of the war to be able to accept the tempting offer to be the first long-term French visitor in the French-Polish mathematical Institute, newly created in Cracow. He had kept an enchanted memory of his stay in Poland, in spite of the allegiance, at that time, of the country to the Soviet Union. I was happy to visit Poland myself. I admired the pretty colored houses reconstructed in Warsaw by the Polish after the destructive bombings during the war. Cracow had suffered less and kept a charm of past times.

An important group of relativists had formed in Warsaw. Leopold Infeld was a Polish Jew who had stayed in Canada, then in the United States, from 1938 and had collaborated for a time with Einstein. He went back to Poland in 1955 following disagreements about nuclear weapons of the United States after the war, also shared by Einstein and Oppenheimer who remained, however, committed to the great nation where they were established. Back in Poland, Infeld became a professor in Warsaw University. With another Polish relativist, Plebanski, he started the creation of a group of relativists. Andrew Trautman, some twenty years younger, was a brilliant element of it. The group has expanded with his students and students of them. Upon recommendation from Trautman, I welcomed, in our team of general relativists, first Richard Kerner, who completed there a thesis of subject proposed by Trautman on Yang-Mills fields, then Piotr Chrusciel, with a post-doc fellowship. Both excellent scientists, they obtained positions in French universities. Richard, though coming from a position as Assistant in a mechanics department, managed the feat of being elected Professor in the physics department of the Pierre and Marie Curie (Paris VI) University. He just became Emeritus after very varied works, many publications including several books and the training of several French or foreign students, among them Brazilians. Piotr was a professor of mathematics for several years in the University of Tours. Then he let himself be tempted by a stay organized by Paul Tod in Oxford University. I had the pleasure of visiting him there and giving a lecture that Roger Penrose came to listen to. Afterwards we had lunch and amiably chatted about science and related subjects. Piotr then left Oxford for a professorial post at the University of Vienna in Austria, nearer to his native Poland and the aging parents of his wife. Piotr was replacing, in Vienna, my friend Peter Aichelburg, who had reached legal retirement age after a rich harvest of interesting works. During my last years of mathematical activity, I collaborated with Piotr Chrusciel on an interesting problem posed by the Einstein equations: research of solutions with data on a characteristic hypersurface. For very complex calculus, we asked the help of a specialist of computational algebraic calculus, José Martin Garcia. Our collaboration was pleasant and fruitful. Unfortunately, José, a Spaniard with a fellowship in France, did not obtain the permanent post that he had wished for in his original country. He has been recruited by a big American society, Wolfram, to write scientific calculus programs.

Prague, Jena and East Berlin

My first trip abroad after the war, organized by the ENS in 1947, was for Prague, with fellow students. We arrived after an uncomfortable trip sitting on a train's wooden benches; it was so at that time. We were warmly received by students of the university. With them we visited this city full of charm. I befriended, particularly, a French-speaking couple. They told me they were not Communists, but were grateful to the Soviets for having liberated them from Nazi occupation. We had promised that we would meet again, but contacts were broken after the Prague Spring.

I crossed Czechoslovakia with Geneviève in my Renault 4L in 1980, to join Jena where a Congress on General Relativity took place. The Prague monuments were then black, like those of Paris before their cleaning by Malraux, and the Charles Bridge was under work. As we were camping, we did not stop in the city, but somewhat farther in a military camp where small vacant wooden houses were kindly proposed to tourists. The following day we crossed the German border. The customs control looked severe. Our back trunk, full to the brim with camping material, gave me some apprehension. A Czech customs officer came towards us, opened and closed the trunk, pretended to inspect minutely some postcards put against the windshield and let us go through the frontier without another formality, comforting my sympathy for the Czechs. After briefly visiting Dresden, regretting that the museum information was in Russian, we arrived in late evening to the hotel of Jena where a room was reserved for us for the next day. The room was free, but the manager sent us back firmly, to the street. We were driving, wondering what to do when the police stopped us: night circulation forbidden! I showed to the policeman my letter of invitation to the congress. He became gracious, phoned I don't know who and brought us to a house belonging, he said, to the mayor. We found there, for free, a comfortable room, and some other congressists. The next day our hotel room was available. All is well that ends well. I don't keep another memory of this congress, which was organized by the competent professor, Schmutzer.

Another congress was organized by a professor in Jena University, Neugebauer. He had the good idea to choose, for it,

Weimar, a beautiful neighboring city that was rich in history. I gave at that congress a conference on general PDE which explains, in particular, why a remarkable property, that I proved for the Einstein equations while studying gravitational waves, explains the non-linear stability of the Minkowski space-time.

In the seventies I went to Brno, then in Czechoslovakia, now the capital of Moravia, part of the Czech Republic. I was invited to a congress on differential geometry by a nice couple — the husband was Slovak and the wife Czech. I liked the small town of Brno. Upon my return, I stopped to visit Bratislava in Slovakia, an old town which looked to me to be somewhat asleep despite its beautiful cathedral.

I was happy, without being surprised, that Czechoslovakia, in contrast to neighboring countries, divided pacifically. I accompanied to Prague, in 2002, my husband Gustave Choquet, who had been elected Doctor Honoris Causa. I was happy, to visit anew, this town which has kept all its human size charm. The internationally-known relativist, Jiri Bijac, had the kindness to ask me to give a conference, which he listened to with attention and commented on fruitfully. He invited me to lunch, where we had an interesting and pleasant conversation.

Vienna

Vienna has possessed an excellent group of theoretical physicists in the Schrödinger Institute, directed until he retired by a great theoretical physicist, Walter Thirring, who was born in 1927 and dead in 2014. Walter was the son of a valuable relativist, Hans Thirring, who gave his name to a general relativistic phenomena, the "Lense — Thirring effect". The original works of Walter concern mainly quantum theories, but his interests were on all theoretical physics. He published many books. A series of them is entitled "Course in Mathematical Physics". In 1977, Walter wished to include in one of these books a chapter on general relativity, of which he was not specialized in. I had the good luck to be chosen by Walter to give a one-month course on this subject at the Schrödinger Institute. I could thus visit, at leisure, the former capital of the Austro-Hungarian Empire, appreciate its lively atmosphere, admire its monuments, especially the Saint Etienne cathedral, and enjoy the wealth of its museums. I had, as students, researchers preparing a thesis, nice and interested, in particular Peter Aichelburg and Robert (called Bobby) Beig who became valuable researchers, obtaining original and interesting results on General Relativity. They were given positions in the Schrödinger Institute or at the Vienna University. Both became my friends. I saw them again with pleasure in congresses related to General Relativity, and of course during my subsequent visits to Vienna. They came to the IHES to celebrate my ninetieth birthday.

I saw again Walter Thirring both in Paris and during my visits to Vienna. On the occasion of a symposium in 1994, he invited me to a concert in I don't remember which church, where his performance on the organ was followed by a cocktail. In 2010, having been invited for a stay in Vienna by Piotr Chrusciel with whom I was writing an article, I saw Walter for the last time. He was then a widower. He gave me the pleasure being invited to dinner, together with the couple Aichelburg, in a restaurant in the neighborhood of the city. Walter came to fetch me from my hotel in a car belonging to him, with a paid driver. We had dinner on a terrace overlooking a beautiful landscape. I am ashamed to say that I don't remember what we talked about, only the kindness of Mme Aichelburg, whom I knew little before that evening, remained. Walter had aged, myself too, of course. He died in 2014 at eighty-seven years old.

Germany

When the generation who had fought us during the war retired, my prejudice against stays in Germany ceased and I willingly accepted stays in institutes that were organizing meetings I found interesting, for example Oberwolfach or the Max Planck Institute on gravitational physics. This Institute directed by Jürgen Ehlers was the first in Munich. I made there an interesting stay, visiting museums and working with Christodoulou, who had, there, a fellowship.

I went with Geneviève to Berlin in 1981, before the fall of the Wall, for a congress of the new association Mathematics and Physics, created by Moshé Flato, where Christodoulou was lecturing. West Berlin was buzzing with activity. We went one Sunday to visit East Berlin, where it was easy to take a return ticket from, by the subway. One arrived to the East by an exit invisible from the outside. What contrast between the two Berlins! Not so much, in fact, to the disadvantage of the East with its beautiful and peaceful alley der Linden. It was difficult to find room in a restaurant, but a sausage seller on the sidewalk allowed us to sustain ourselves. We had some difficulty finding our way back to the subway entrance to go back to West. A passer-by questioned told us that such an entry did not exist. Fortunately, another took us there.

After the fall of the Berlin wall, the Munich Max Planck Institute, still directed by Ehlers, was displaced to the neighborhoods of Berlin, first Potsdam, then Golm. I went to Potsdam, a pretty city with a historical castle, as I was invited by an excellent, original and rigorous specialist of mathematical problems posed by the Einstein equations, Helmut Friedrich. I had made a friend of Helmut during a common stay at the Les Houches School in 1982. Golm is less interesting than Potsdam for tourism, but I went there several times with interest and pleasure, working with other visitors, in particular, Piotr Chrusciel. A visit in 2006 was on the invitation of the director succeeding Ehlers, the great mathematician Gerhard Huisken, who was also the author of a difficult proof of a Penrose conjecture and an exceptional lecturer. From Golm, after passing through Berlin which was full of new buildings, I went to Poznan to celebrate the hundredth anniversary of the birth of Jean Leray that was organized by his former student, the active mathematician Granas. My last stay in Berlin was in 2015, to speak about my encounters with Einstein during the congress organized by Herman Nicolai to celebrate the hundredth anniversary of general relativity. It leaves me happy memories of meetings with old friends.

I also keep a pleasant memory of a visit in full winter to Tübingen, white with snow, on the invitation of Professor Pfister.

Remo Ruffini and the Marcel Grossman Congresses

Remo Ruffini is Italian, almost French since he has roots in Brigue where he is born. He speaks perfect French. Remo is a theoretical physicist. After working on mathematical problems in General Relativity under the aegis of Ferrarese, he sojourned in Princeton in the laboratory of J. A. Wheeler. He contributed in an essential way, with his student Christodoulou, to the first studies on black holes. Since 1978, he has been a professor in the physics department of the University La Sapienza in Rome. He has turned his research focus to astrophysics, a domain in full expansion. These last years he specialized in “gamma ray bursts” — these surprising explosions of enormous energy which happen pretty much everywhere in the universe and have an origin still containing many mysteries. Remo is a remarkable man, with an exceptionally warm personality full of energy to undertake and materialize. While continuing his personal research on astrophysics, he worked on developing international collaborations. He created the ICRA, "International Center for Relativistic Astropysics", located now in Pescara (Italy). He was the initiator, then the organizer, of the "Marcel Grossman meetings on General Relativity" that meet every three years. Marcel Grossman was a mathematician of Hungarian origin who had been a friend of Einstein in Zurich and had helped him to find the geometrical formulation of General Relativity.

I had the honor of being invited by Remo Ruffini to the first Marcel Grossman Congress he organized in the Theoretical Physics International Center in Trieste in 1975, with the collaboration of its director, Abdus Salam. I take pleasure today in looking at the list of participants. Many names remind me of friendly colleagues. I remember a lunch with Abdus Salam, a man famous for his discovery with Weinberg of unification of electromagnetism and weak interaction. Salam was, like Einstein, a kind man, simple but radiant with intelligence. I read later an autobiography that he has written. He was a convinced Muslim, but preached the Christian love of neighbors.

Remo Ruffini kindly chose me to represent France in the international committee sponsoring the Marcel Grossman meetings that he continues to organize pretty much everywhere in the world. I went to many of them, always with interest and pleasure. At the fourteenth congress in Rome, from 12 to 19 July 2015, there were 1,200 participants. Remo offered me an escort: it was my daughter Michelle. We had a most agreeable stay, meeting with old friends and hearing scientific news, in this city rich in history.

13

ACADEMICIAN 1979

The Academie

My master and friend, Jean Leray, was a member of the French Académie des Sciences since 1953. He was a conscientious Academician, attending all meetings when he had no other unavoidable obligation. He had worked for the reform of this venerable institution, wishing for its rejuvenation through the vote of a new law for the elections: among the newly elected members there should now be a certain number of young personalities, that is, those less than fifty-five years old. It is true that without this rule, because of the longer life expectancy, the Académie would soon be composed only of decrepit old members. In 1976, Leray suggested to me to present my candidacy; it was the procedure at the time. Lichnerowicz, who had told me in the past, "You will be the first woman in the Académie", approved this suggestion. This required that I write a "notice", i.e. a report about my scientific works. Leray told me, "You will see, it is very pleasant." He was right, I took pleasure in explaining the motivations that inspired my researches and their results. I brought my report to Leray, who approved it, but said, "This time I will support Lions (Jacques Louis), but you should take date, I will support you next time." The idea of a candidacy to the Académie was not mine, in fact I never looked for honors, I more appreciate affection or only sympathy from my "human brothers" as said the poet Villon. However, the idea of being a candidate without the prospect of being elected deeply displeased me, and I gave up this candidacy. Lions was elected without it upsetting me because I had for him friendly feelings and appreciation as researcher and organizer.

Leray had told me in 1976, "You made a blunder." This did not prevent him from having me elected in 1978 as corresponding Académie member, an honorific title which did not require candidacy and is now suppressed. At the following elections of members in 1979, Leray advised me again to be candidate, saying that this time he would support me. He told it to my colleague and friend, Maurice Roseau, while suggesting to him to also be candidate, as he had done for me in 1976. Roseau, a few years younger than me, accepted while kindly telling me he knew he would not pass before me. Roseau was elected at the following session in 1982. Leray had written the report and asked to the new Academician I was to present it, I had done it willingly. Roseau learned, only years later, that it was Leray who had written this report, he had held towards me an undeserved gratitude.

Visits

Returning to 1979, I had my notice ready, but I had to make the visits to Académie members, required at the time. In fact, while I had kept a bad memory of visits made when I was candidate to the Sorbonne, I found pleasant those done on the occasion of my candidacy to the Académie, perhaps partly because it was important for me to obtain a post in Paris, while the Académie was a luxury with nothing essential for me. Some Academicians, for instance Robert Dautray and François Jacob, told me that the support of Jean Leray was for them sufficient recommendation, others gave me an interview in the Académie or in their laboratory.

The first importance to be elected in the Académie is the vote of the section where you are candidate. In my case, it was the mechanics section to which belonged Leray, Lichnerowicz and Lions who favored me. An influential member was Maurice Roy, director of ONERA. (National Aeronautic Research). My kind of mechanics didn't interest him, but he was a friend of my friend Jacques Valensi, a good point for me. When I met him he said, "You don't do quantum mechanics, I hope?" Reassured on this point he promised me his vote. I had been told, and it was verified, that Legendre always voted like Roy. Another important member of that section was Paul Germain. I called him; though I was not a "true mechanician", he could hardly snub me. His escape was to answer, "What I want is that Malavard be elected." Malavard, a "true mechanician", was candidate to the other post, without age condition, open in the section (mine was reserved to "youths"). I asked Germain, "Do you want me to visit you?" He answered, "No of course." I had known Favre in Marseille; I was on good terms with him despite his disagreement with Valensi. Brun did not know I had fled his course some thirty years before. My father had been his colleague in the Sorbonne, I don't know where his vote went, nor if he came to vote because he died that year. A retrospectively amusing memory is from the visit I made to Darrieux, already quite old, in a faraway suburb. He lived with his daughter, who insisted on offering me a cup of tea before I left, as an excuse on behalf of her father to have taken me a whole afternoon to speak of his works (which I have forgotten), while too old to come: he would not participate in the election! One day, standing at the entrance of the assembly room, thinking, "What good is it to go through this hassle to be in the Académie?" Jacques Louis Lions, passing near me, interpreted my look as worry and told me, "Don't worry, you will be elected." Understanding better my feelings from my disillusioned pout, he added to comfort me, "It is paid." It is not the hope of this small sum that cheered me up but the kindness of Jacques Louis. He was a true friend. I was pained by his untimely death.

I do not keep precise memories of meetings with members of the mathematics section. I already knew them all. Néron, Thom and Serre were friends, having been with me in the ENS. I had good relations with Garnier and Henri Cartan. Dieudonné told me, "Mathematicians will vote for you." I was not sure of Schwartz because of his tense relations with Leray. Meeting him at the entrance of the assembly room, I said, "I did not think of the Académie, but now that I am candidate I wish to be elected"; he answered, "Of course."

I knew also, at least by name, most members of the physics or sciences of the universe sections. Many had been colleagues or students of my father. I don't remember what they said of my candidacy, but I recall my pleasant meetings in the Académie with, among others, Castaing, Friedel, Grivet, and also Guinier, who gave me his support for missions abroad when he was responsible for them. I don't remember if I made a visit to Alfred Kastler. On the other hand, I recall that when I was an Academician, my respect for him grew when he presented for the "Comptes Rendus de l'Académie" a paper on general relativity that was sent to me for referee. This paper was not false in itself, but was not in agreement with the definitions of Einstein's theory. Without refusing the Note, I made this comment in my report, in principle anonymous. I received a letter (not anonymous!) of Kastler saying that he withdrew his paper but wished to have more information about the content of general relativity. He seemed satisfied by the explanations I sent him. I was happy to get to know, in the Académie, other remarkable members of the physics section. Anatole Abragam was a very intelligent man, with caustic wit, but I only had good relationships with him. I am grateful to him that in his autobiography, in which he is not sweet to professors who taught physics in his youth, he writes, "Unfortunately Georges Bruhat has not been my professor." Louis Néel, born in 1904, had probably not followed teaching from my father, but he was a powerful mind, a man with a considerable presence, who honored me with his sympathy. I regretted when he stopped coming to the Académie to go to live in Brive la Gaillarde, perhaps with a daughter, until he was ninety-six.

Among the doctors, Hamburger, whose book "Reason and Passion" I much appreciate, was very friendly when we first met and thereafter. We were crossing together the institute courtyard, with a third person, I forgot who, when Hamburger said, "Next time we will elect the little Le Douarin." I was not shocked by this friendly familiarity concerning the sex called weak. Certainly, Nicole Le Douarin is not a weak woman but a biologist of exceptional talent, and a praiseworthy leader that all Academicians could appreciate when she was "perpetual secretary." She is a remarkably open and intelligent woman. I appreciate her friendship.

Jean Bernard had sent me a kind word to tell me not to bother with a visit. Our relations were less familiar than with Hamburger. However I learned to appreciate this great doctor as he deserved. While he was a student in Ecole Centrale my son Daniel had a lump in the groin. The school's doctor wanted him to have a biopsy. I called Hamburger who said, "I am retired, but Jean Bernard still practices, call him." I did. Jean Bernard asked Daniel to come to his home, and after interrogation and reflection gave for advice to Daniel to go for three weeks in the mountains, skiing if he wanted. Daniel followed that prescription and the lump disappeared. As I was warmly thanking Jean Bernard, he said, "You know, we took a risk." He was a great doctor. I did not hesitate to follow his advice a few years later when he supported the candidacy of Maurice Tubiana, Gustave Roussy Institute's director. I am glad for the election of Tubiana whom I met, always with pleasure, in the Académie. Reading the article about him in Wikipedia, my natural modesty is surprised at the amiability this great man has always shown to me. I am grateful to him for his help to my friend Vladimir Georgiev, a Bulgarian mathematician, a known specialist of partial differential equations. He had been diagnosed with cancer. Fearing that the prescribed treatment was not the right one, Vladimir wrote to me for advice about the name of a doctor he would come to see in Paris. I called Tubiana, he suggested that Vladimir send him his medical file. Having received it, Tubiana wrote to Vladimir and called me, approving the correctness of the prescribed treatment. Vladimir healed. His interesting mathematical works earned him a professorship in Pisa University; he invited me for a month in the excellent university of this beautiful city full of architectural treasures. I enjoyed this undeserved reward. It is Tubiana who gave me, in 2010, the decoration of "Grand Officier de la Légion d'Honneur" due to my works on mathematical physics and, I guess, the recent measures on parity. Tubiana came to the Académie especially for this formality and gave me the ritual hug in the anteroom of the Assembly room.

Another doctor in the Académie in 1979 was Jean Dausset, famous for having performed the first kidney transplant. I remember, with some satisfaction, my brief conversation with him as we were sitting next to each other at the Académie; he was looking with contempt at my paper whose content I had tried to explain to him. He said, "What is the use of all that?" I answered, "Nothing, sir" and left without further comments.

Most members of the second division were more friendly than Dausset. I keep a particularly good memory from my visit to some of them in their laboratories where they spoke to me about their own works and expressed sympathy. For example, P. P. Grassé made before me lively and interesting statements about his beliefs which one finds in his book "Toi, ce petit Dieu." He was very kind to me, saying that I seemed to him worthy of being the first woman in the Académie des Sciences. The naturalist Cauderon, who was later perpetual secretary of the Académie d'agriculture, received me warmly and convinced me easily of the importance of his works.

Election

Lichnerowicz, in principle, did my report, but he was in the US. Therefore it was Leray who presented it; he told me he had improved it a lot. He also revealed to me that the delegates of the Mathematics (Henri Cartan) and Sciences of the Universe (Jean-Claude Pecker) sections intervened to say that their sections were favorable towards me. I knew Pecker well. He was admitted a year before me in the ENS but a student with me because of the anti-Semitic persecutions during the war. He is a friend. I was slightly displeased when he revealed to me later that he had voted for Roseau, whom he knew well, because he was afraid he would have no vote! I was elected with a comfortable majority, but, and that's good, Roseau did not have only Pecker's vote.

The Académie's president, the naturalist Gautheret, accompanied by Leray, came to our home to announce my success, the first woman to be elected in the Académie des Sciences (Branly had been preferred to Marie Curie). Gustave offered champagne. He was himself an Academician since the previous session. He showed unreserved satisfaction of my election. I must do justice to my husband that he never put obstacle to my work. He had such a veneration for mathematics that it never occurred to him to oppose their practice. He was not jealous of my eventual success because, fortunately for me, he was rightly convinced of his superiority. Later when I progressed in "Légion d'Honneur" while he trampled as officer, he said without rancor, "It is because she is a woman" which is true in our era of parity. Gustave, in an interview published in the number 62 of the newspaper "Tangente" says that mathematicians are of two categories, strategists and tacticians; the first, to which he says he belongs, being obviously superior. For me, mathematics are a tool; I acknowledge this being rather a tactician. Gustave knew to be superior as mathematician, and I had no desire to challenge that. Unfortunately he was convinced that this superiority extended to all domains, even the most trivial; the resulting criticisms were sometimes hard to bear for those around him. However I am grateful to Gustave for facilitating my scientific research work, something that the meanest men would have hindered.

Back from the States, Lichnerowicz organized, in the Collège de France, a little reception for the celebration, by our team, of my election in the Académie.

14

LIFE CONTINUES 1979–1990

Teaching

During the years 1979–1990, I continued to divide my teaching between undergraduate classes in mathematics and mechanics departments, which were now called UER (Unité Enseignement Recherche). Neither of these two UER wished to let me give a graduate course concerning mathematical physics. I would have liked to teach students of the mechanics UER the foundations of quantum mechanics, because to teach is the best way to learn, but my colleagues refused forcefully. I continued for a few years to teach distributions theory as a mathematics' option, but a professor of that UER, tempted by this course, asked that I leave it to him. In compensation the mathematicians allowed me to teach an option where I initiated a small number of students to general relativity. This subject had been suppressed from the physics UER after Marie Antoinette Tonnelat left its teaching for that of philosophy of sciences to literaries. My course on general relativity pleased me enough, because it had interested listeners, among them physics students, but it was interrupted by the administration after a few years. Students wishing to learn about Einstein theory who came from a physics UER were not counted as students by mathematicians. One year, when the recognized students did not reach the prescribed number (eleven I believe), the students from the mathematics UER got a notice after several weeks that the course was suppressed, without the other students, nor myself, being told. When I arrived to give my course, some of the students waiting for me belonged to the mathematics UER who had come to reproach me for this unexpected suppression! They regretted it more than me, though not sorry to be exempted from work, they were shocked by the process followed by administration that had thought unnecessary to contact the professor responsible.

One year I accepted a beginners' undergraduate course that no colleague wanted to undertake. At the first session the background chatter in the amphitheater of two hundred students was unbearable. I went out as I had done in Marseille, saying that I would deliver my course when a delegation of students would assure me of their attention. The delegation came, incidentally saying they did not understand why I was complaining, the noise being much worse in my colleagues' classes. I resumed my course, but I refused to take again that teaching the following year, comforted in this decision by the boorishness of colleagues who had gotten rid of a mentally ill assistant by dumping him on me. This guy would obtain from his doctor a three weeks' break, did not give advance notice that he would not come to the university, returned without warning either, had the janitor see that no student was present, and take another three weeks' break. I was forced to do his work, at least partly, in addition to mine. I called him in my office. He told me it was too hard for him to remain for a whole hour in front of other people. I said I understood him but advised him to take an extended leave that would allow the administration to replace him. I obtained the name of his doctor and explained to him the situation. The assistant took a long leave, I did not see him again. But when I reported the facts to a colleague, he told me that the behavior of this assistant was well known, that is why he had been gotten rid of him in attributing him to me!

My teaching in the mechanics UER did not lack listeners and they were peaceful, but I saw their previously acquired knowledge lower a little more each year as well as their results at the final examination, though the level of my course was increasingly lowered.

All these facts explain why I did not ask for the prolongation of the three years of teaching allowed to university professors after the legal age of retirement, at sixty-five years old (sixty-six for a parent, like me, of three children).

My last years teaching mechanics were not very rewarding, but I was fortunate to be assisted by a "maître de conférences" (associate professor), Mantion, with whom I had friendly relations. He was both dedicated and highly competent.

My Children

Daniel had been a brilliant student in Math-elem (scientifically oriented last year of high school). He was undoubtedly a scientist, destined to be a researcher like his parents. He was not clear on what discipline would be his choice. As a child, though rather turbulent and liking sports, he spent long hours assembling beautiful models. He had asked, for a gift, "The Little Chemist" and made, in the basement, experiments giving us some concern. As soon as the first computers appeared we had to buy one for him. The natural result for him was therefore, after the baccalaureate (exam at the end of high school), to prepare for entrance in a scientific "Grande école", and to register in the best Hypo-Taupe present on the market, that is, in 1979, in the lycée

Saint Louis. The director of Saint Louis was the brother of a colleague and friend. As a good mother ready to beg a favor for her son, I called him. This honest man replied very curtly that he was examining Daniel's records, impling at once his admission to this prestigious lycée. In fact, it was not a good thing. The pressure prevailing in the Taupe of Saint Louis was not suitable for Daniel. I understood him because I myself had not liked the competitive atmosphere in the Taupe at Fénelon. But Saint Louis was worse and Daniel may be more sensitive; he had shingles handicapping him for several weeks. For that reason or another, he was received the first year only in "École Centrale". He did not wish to try another time. His father and I let him be free of his choice. We were happy with it, since Daniel is now a great biologist and member of the Académie des Sciences.

Geneviève passed without problem a scientific baccalaureate, with good marks in mathematics and physics, but her more varied interests did not encourage her to specialize in these fields. Perhaps influenced by my old unsatisfied wish, she directed herself towards a medical career. She completed, with no break, all the stages of these long studies without, apparently, the important amount of work they are supposed to require. She was admitted to the internship at the three places where she had applied, with a very good rank in Lyon, but unfortunately (for me), a less good rank in Paris. She took Lyon, she said, to be able to choose the specialty she wanted, dermatology. After her medical thesis, Geneviève did a scientific thesis, and then accepted a research position in oncology at the big private Institute Bio-Mérieux. She remained there for more than ten years before returning to the practice of dermatology, seeking new skills in this discipline. Despite her devotion to her three sons, she still finds time for sculpture, trying new techniques.

My eldest daughter, Michelle, was now living in France. She had a son, Raphaël, born in 1977 and a daughter, Prisca, born in 1980. I loved my grandchildren and was happy to be with them. During the years 1984 to 1988, I was lucky they lived in Paris' suburbs, in Ermont. Often they took the train alone to come spend the week-end in my home in Antony. Gustave liked children; he has never put up any objection to their visit, even when their traditional Disney Channel show on Saturday evening prevented him from watching another movie. He was, for my grandchildren, a better grandfather than the true one. On Saturdays or Sundays I took these children to some garden with attractions of their age. These were not lacking in the Paris region. On sunny days, we sometimes went to Dammartin. Raphaël keeps good memories of his stays in the big family house and his meetings there with his Great Uncle François. The youngest daughter of Michelle, Abigail, was born only on December 14, 1987. I knew her well only later, in spite of my visits in Guadeloupe where Michelle stayed from 1988 to 1997.

More Mathematics, Collaborations with Christodoulou

In February 1978, I was invited to Torino for a month by Mauro Francaviglia. Demetrios Christodoulou was staying there at the same time as me. It was his wish because he wanted to devote himself more to mathematics than before. Demetrios had been a brilliant student in the schools of his native Greece. At sixteen, wishing to go towards research in mathematical physics, he came to Paris with his mother to seek advice from another Greek, Achille Papapetrou, who advised him to continue his studies in the States and see J. A. Wheeler. It was wise advice because the American school curriculum is more flexible than the rigid French one. Wheeler greeted Demetrios kindly; after a quick schooling in Princeton University he sent him to Remo Ruffini, then fellow at Princeton, to prepare for a doctorate. Remo was delighted by his student; he told me, "I have an extraordinary student, he solves all the problems I give him!" Demetrios acquired, thanks to Remo and John Archibald, a useful physical intuition, but the lack of generality and mathematical rigor of some results left him unsatisfied. After his thesis, Demetrios, tired of the States, had taken a position as a researcher in the group of general relativity of the Max Planck Institute in Munich, directed by Jürgen Ehlers. He came to spend some time with me in Torino to give more mathematical body to his works. John Archibald told me later that he had been a bad mentor for Demetrios; I think he did not do himself justice. I believe that the brilliant scientific career of Christodoulou is due, besides his remarkable and fertile imagination, to the conjunction of his cultures in physics and mathematics. In 1978, I was near fifty-five and Demetrios some twenty-five years less, however we quickly became friends. Demetrios said he could hardly believe that he was younger than my oldest daughter and added warmly, "We will always be friends." It was a just prediction, despite our geographical distance. Although he is not a lover of travels and congresses, only a serious accident of his mother prevented him from coming, as planned, to the celebration at the IHES of my ninetieth birthday.

Going back to Torino in 1978, under the impulse of Christodoulou and with his active collaboration, Francaviglia and I took anew, with a fresh energy, some mathematical problems of General Relativity. Demetrios studied, in one night, the book of Adams' on Sobolev spaces. We attacked three subjects: geometric formulation of the evolution problem for the Einstein equations, existence of solutions with a given behavior at past infinity for hyperbolic semi-linear equations, and global existence of harmonic coordinates. The three of us discussed, but made little progress. Mauro was too much of a perfectionist. I suggested each of us take charge of the writing of one of the subjects, that would be published under our three names, it was accepted. I took the first subject, the fairly quickly finished draft was submitted and accepted by "Annales de l'Institut Poincaré". Demetrios took the most original, the second one, and he made a very synthetic writing of his solution. We sent it to Bob Geroch, responsible for the "Communications in Mathematical Physics". Bob refused the publication with the comment, "There are too many hypotheses, though they are probably necessary." We then sent this text to "Annales de l'Institut Poincaré", and it was accepted without discussion. I had looked at the problem and sketched a demonstration, less elegant but easier to follow, of its solution. I exposed its development at the congress organized in MIT to honor the sixty-five years of Irving Segal, himself the author of pioneer work on the global existence of solutions of semi-linear wave equations. My work, co-signed with Christodoulou, appeared in the conference proceedings. I proposed its generalization to the Einstein equations to a student of Francis Cagnac in Yaounde, Norbert Noutchegueme. He drew from it a very respectable doctoral thesis. Mauro never wrote the solution of the third subject, which was nevertheless interesting.

Wishing to continue our collaboration, Demetrios had me invited for some time in Munich where he was a researcher at the Max Planck Institute, in the unit on gravitation directed by Jürgen Ehlers. It was a pleasant stay with new and old friends, in a modern city with old museums. My scientific life there has been busy. Demetrios was full of ideas, which required serious work to work out. Two of them led to articles published in excellent journals. The first improved the properties of weighted Sobolev spaces; it allowed the extension of results that I had obtained with Alice Chaljub Simon for the solution of Eisteinian constraints. Demetrios and I discussed it in Munich. I completed the result and wrote the article upon my return to France. Demetrios came on his own initiative to bring me his help, and was a bit disappointed that I had not needed it. Demetrios had another original idea: to use the conformal transformation of Minkovski space-time on a bounded domain of the Einstein cylinder, used by Penrose to study asymptotic behaviors, to prove global existence theorems for small initial data of solutions on Minkowski space-time of the Yang-Mills equations, these being conformally invariant. I found this idea excellent, but I thought the merit should be left to this young researcher who was just beginning his career, and he agreed with me. However, captivated by another interesting problem, the Yang-Mills charge, he neglected to develop his idea, and it lost its interest a few years later when Eardley and Moncrief, in a remarkable work based on a different idea, proved this global existence without restriction to small initial data. In the meantime I had been interested by spinors and the Maxwell-Dirac equations. The Eardley-Moncrief method did not apply to them, but I had found a definition of their conformal transformation and applied to them the method imagined by Christodoulou. I suggested to Demetrios to write, at once, his proof for Yang-Mills equations as a note for "Compte — Rendus de l'Académie". It would appear quickly and we would cosign an article proving the global existence of solutions of the standard model, Yang-Mills, spinors, scalar field. This was done; I wrote the article but Demetrios added a useful remark on the asymptotic behavior of solutions. Our article was published in the "Annales de l'École Normale Supérieure", as he had wished.

Christodoulou's talent earned him, on Ehlers' proposal, a Humboldt fellowship to spend a year abroad. He wished to spend that year at the Courant Institute of New York, host of remarkable specialists on partial differential equations. I recommended him to one of them, Louis Nirenberg, whom I admired and befriended, and he was admitted. Demetrios found, at the Courant Institute, a former student of Nirenberg, Sergiu Klainerman, a brilliant researcher who had proven a global existence theorem for equations of the type of those of general relativity in harmonic coordinates, but in at least six dimensions. Sergiu had come to see me in my office to ask if this result had a physical interest. I announced to him the impending arrival of Demetrios and his possible collaboration. The possibility became a reality. It led to a result that made them famous a few years later, the resolution in the negative of a conjecture made by Einstein many years before: the identity with Minkowski space-time of any complete and vacuum Einsteinian space-time.

At the end of his Humboldt fellowship, Christodoulou, at Ehlers' discontent, wished to remain in the United States, preferably in Syracuse where a young relativist, very active and open to new ideas, Abhay Ashtekar, was staying. The department's head, Joshua Goldberg, was known to me and friendly with me. My recommendation letter was probably unnecessary, but I wrote it willingly and Demetrios went to spend a few years in Syracuse. I went, for a few days, one winter. It was very pleasant in spite of the cold, thanks to Abhay and Demetrios. Christodoulou went back to the Courant Institute as a professor before joining Sergiu in Princeton, where they completed together the important theorem I had quoted before.

Christodolou left Princeton for ETH in Zurich, closer to his native Greece where he makes long stays without interrupting his scientific activity, always renewed.

MIT Collaboration with Irving Segal

In 1979, I renewed my old friendship with Irving Segal at his sixtieth birthday, celebrated in MIT (Massachussetts Institute of Technology) where he was a professor. Our interests in mathematics and physics were still close. Irving invited me, in 1981, to work with him in MIT for a month. He directed with great dedication the researches of several students, mathematicians interested in physics who shared his belief in his cosmology, an unchanging universe, invariant under the Lorentz group. Irving was then divorced, remaining very attached to his three children. He had kept the big house, called "Moon Hill", that he owned in Cambridge and rented out a small apartment at the garden level. It was free when I came and I willingly accepted to rent it for the month. Irving was interested in the method used by Demetrios and myself to prove the global existence of solutions to the Yang-Mills equations on Minkowski spacetime. He hoped we could extend this result to his cosmos. We attacked this problem and a combination of methods and ideas allowed us to achieve the result. Irving entrusted the drafting of the demonstration to his student and favorite disciple, Paneitz. The latter performed his task, using the definitions and specific notations of his master, which were for me, I have to admit, foreign. The article cosigned by the three of us appeared in the journal dear to Irving, "Journal of Functional Analysis", thirty years after our first article in collaboration written in Princeton. I, however, granted myself the satisfaction of writing a demonstration with my usual notations for a note in the "Comptes Rendus de l'Académie". Unfortunately a tragedy occurred soon after: Paneitz, a tall, apparently robust young man, drowned while trying to cross, with Irving who was a man of about sixty and not particularly athletic, a lake close to a place where both had participated in a congress. Irving told me this later, after having consulted a psychologist, a very American reaction, to alleviate the sorrow caused to him by this death. Irving invited me again in 1983, for three months, but I could only stay six weeks because Gustave wished to make a long stay in Cameroon and we had a policy of never leaving our children without at least one of their parents. I had, again with Irving, fruitful discussions on mathematics, and I listened with interest, but without being convinced, to Irving's arguments in favor of his chronogeometry, supported by a detailed analysis of observational data. At the end of my stay, Irving and I went to New York to participate in the dinner to honor the eighty years of the famous mathematician K. O. Friedrichs. K. O. told me very kindly that he would now come seldomly to the university, but would do so to meet me if I had an occasion to come to New York. Unfortunately he died the following year. That night I slept, as he had offered, in the apartment of my friend Englebert Schücking and chatted pleasantly with his wife, who was courageous and kind in spite of serious health problems.

In 1986, I made another visit to Irving Segal. I had been invited by the association of women mathematicians to give, in New Orleans, the conference they organized every three years to commemorate the memory of Emmy Noether. For financial reasons I had to visit, upon this occasion, two other American universities. The organizers had planned for me, with Irving Segal and Cécile DeWitt, stays in Cambridge (USA) and Austin. Irving was in New Orleans at the same time as me. He took me to a jazz show, a specialty he loved. He told me it would be criminal not to enjoy it when in Louisiana. I appreciated the music, but with less enthusiasm than some other listeners.

I was invited again in MIT, to give a conference at a congress dedicated to Von Neumann. Irving then told me he was remarried with a woman some thirty years younger than him. That year, I was accommodated in a nice room of the apartment belonging to the former wife of the famous mathematician John Tate. I had the occasion to appreciate the young wife of Segal during an excellent dinner in their house of Moon Hill. They had a pretty little girl, about four years old. She came for a walk with Irving and me during my last visit to Cambridge. Her mother had undertaken with success medical studies; the father took much care of the child. It was perhaps too tiring for him: he died suddenly of an infarctus in 1998, a little before his eightieth year.

During my visits to MIT I could admire the skyscraper built in Cambridge next to the river and visit the rich museums of Harvard University and the city of Boston. I had the pleasure of dining with my friends, Georges and Alice Mackey, and chatting with them again.

"Analysis, Manifolds and Physics" I and II

Contrary to other mathematicians I always liked to write, perhaps a gene inherited from Georges Bruhat, my father.

In 1982, Cécile DeWitt and I revised our book, "Analysis, Manifolds and Physics", without the collaboration of the third author, who was occupied by other tasks. We completed it substantially, with a chapter on connections seen in a more geometric light than before. In 1989 we wrote a second book containing, in problems form with detailed solutions, important applications of the theories treated in the first. These two books, republished many times, exist now in kindle formats.

Travel and Work in Italy

My friend Carlo Cattaneo had planned to organize, in 1982, a CIME (centre italien de mathématiques estivales: Italian center of summer mathematics) course. I had accepted, with pleasure, to give a course on wave propagation. At his urging request (by phone), I had accepted the advancement of this date to the summer of 1981. Carlo knew that cancer was making his life expectancy short. In fact, he did not even live until then. His very competent student, Giorgio Ferrarese, took charge of the course planned by Cattaneo. It took place in the CIME of Brixen. I arrived in this beautiful Dolomite city from Rome in Giorgio's car. I recognize myself with difficulty in the photo he took then, which he offered to me in a beautiful silver frame at my ninetieth birthday. Despite the regret we all felt at the absence of Cattaneo, the stay in Brixen was pleasant. I had interesting discussions with Guy Boillat, a deep and original researcher, and with Antonio Greco, who was already my friend. I met Tommaso Ruggeri, who has since then become a collaborator and a friend. Antonio and Tommaso already knew each other. They decided to ask, for me, an invitation from the CNR (Italian National Research Center) that they would share equally. It is how in 1982 I spent fifteen days first in Bologna with Ruggeri, then another fifteen days in Palermo with Greco. Fortunately I had two research topics for which their collaboration was welcome. I had recently introduced a gauge, which I called temporal, to solve the Cauchy problem for the Yang-Mills equations on a curved space-time. Tommaso told me, at once that he knew general relativity and computations linked with it and was quite ready to look, if it was possible, to find an analogous gauge for the Einstein equations. We therefore undertook together the rather complicated computations which led to the result, the first local existence theorem which does not use space-time harmonic coordinates, but the data of the shift, that is the slope of the time line. In the work with Ruggeri, we chose, to simplify computations, a zero shift, that is, time lines perpendicular to space sections. Tommaso took me for some beautiful walks in the picturesque surroundings of Bologna and we became friends for life. My work with Ruggeri, summarized in 1982 in "Comptes Rendus de l'Académie", was published in 1983 in "Communications in Mathematical Physics" with the referee's (I think Jerry Marsden) comment, "It is amusing." Later, when I was visiting Chapel Hill in 1996, Jimmy York suggested we extend, together, the result to an arbitrary shift. We did it with the collaboration of two students of Jimmy's, Andrew Abraham and Arlen Anderson. This new method to treat the Cauchy problem has sometimes been used by relativists, in particular numerical analysts. Ruggeri's cooperation was not forgotten.

Arriving in Palermo for the second half of my Italian CNR stay in 1982, I embarked, with the collaboration of Antonio Greco who already knew the subject well, on the construction of high frequency asymptotic waves for the standard model coupled with Einstein equations. Our work was published in 1983 in the "Journal of Mathematical Physics".

Les Houches' Summer School

I mentioned before the summer school of theoretical physics created in Les Houches by Cécile DeWitt. All physicists who stayed there, that is almost all theoreticians, keep from it a marvelous memory. In 1982, Cécile wished to have an exceptional session on general relativity and asked my advice for the choice of an organizer. I named Nathalie Deruelle and Thibault Damour, two excellent young relativists. Cécile preferred to entrust the organization of the session to Nathalie, telling me she trusted a woman more for this kind of task. The session was indeed quite a success. Thibault gave a very interesting course, I gave a few lessons. I was happy to get to know Nathalie better, an author of interesting works and, in collaboration with Philippe Uzan, a very good book on general relativity. She became a friend that I am always happy to meet.

Supergravities

In 1983, the supergravity theory, initiated by Freedman, Ferrara and Van Niewvenhausen, was developed independently by a collaboration Deser — Zumino. This theory aims at integrating, in the equations, quantities whose product does not commute (as ordinary numbers) but anti-commutes, that is, changes sign by permuting factors. The idea was that, in addition to the typical particles, there exist "superparticles" that obey such laws. Their existence would erase some difficulties appearing in a unified theory of gravitation and the other fundamental interactions, at the quantum scale. The corresponding classical equations had given birth to some publications that seemed to me to be of doubtful rigor. I therefore undertook to search for a precise formulation of these equations and a rigorous proof of the existence of solutions. My results awoke the interest of Moshé Flato. He asked me to publish them in the journal he had just created, "Letters in Mathematical Physics". My work on the famous supergravity theory in dimension 11 that became, I think, interesting with string theory, has been published in "Communications in Mathematical Physics" in 1985. Another work on this super-gravity, done with the collaboration of D. Bao, J. Isenberg and P. B. Yasskin, was published the same year in "Journal in Mathematical Physics".

I had the pleasure of being invited in the '80s to several meetings touching on differential geometry or mathematical physics. I could describe the rigorous results I obtained in supergravity theories, and write them up for the acts of these congresses. The first of them, "Journées Relativistes 1984", published by Springer, was organized, in the building belonging to the CNRS at Aussois, by Philippe Tourrenc, then the director of the theoretical physics team of the Poincaré Institute. Philippe liked my talk; since then he has given me his sympathy and support. I was happy to have in Aussois the company of my Chinese friend Hu Hesheng.

Interesting and pleasant opportunities to exhibit and write on other aspects of my works on supergravities were: first in Nagpur, India, at a symposium organized to commemorate the centenary of Sir Arthur Edington; then in Oregon at the colloquium "Global problems in General Relativity" organized by Professor Flaherty. My lecture interested Robert Bartnik, a young Australian mathematician-relativist who was already an author of valuable work; it earned me an invitation to Canberra in 1988. A "Geometrodynamics" congress organized by Professor Prastaro led me to a pleasant stay in Cosenza and a magnificent journey along the Calabrese coast until Salerno, driven by a young geometer whose name, how ungrateful of me, I forgot. We had to delay our leave for twenty-four hours, the religion of our colleague and friend Sternberg forbidding him to travel on a Saturday. I had been happy to meet this great geometer, I appreciated his book which appealed often to intuition without neglecting rigor. I had expressed my astonishment at his assertion that we would in the future be able to communicate with the dead, and had not found an answer to his justification, "We now know what the sun is made of, though we never went there."

Another pleasant and interesting colloquium, "Differential Geometric methods in Physics", organized in Salamanca by Perez Rendon, a professor at this university famous since Antiquity, gave me the opportunity of a conference, "Mathematics for Classical Supergravities". The lectures were given in the old beautiful building of the university and the lecturers were accommodated in a castle formerly belonging to the Kings of Spain, located in a beautiful nearby park. The well known Italian geometer Benenti complained that the perfect nocturnal silence prevented him from sleeping.

Each of my lectures on supergravities has been included in one of the volumes published by Springer or World Scientific. I also wrote chapters for books in the honor of scientific personalities, in particular Vaidya or Ivor Robinson.

In June 1987, I was invited, on the recommendation of G. Marmo, I think, to give a course in Ferrara (Italy) for the spring school "Geometrical Methods in Theoretical Physics", organized by G. Morandi. My course, "Graded bundles and super manifolds" has been published as a small mathematics book of about one hundred pages by Bibliopolis.

My works on supermanifolds and supergravity have interested mathematicians. However, supergravity did not lead to the desired goal: quantization of a unified theory of gravitation and other interactions. Moreover, no supersymmetric particle has been observed. As for me, being a tactician rather than a strategist, I did not engage into speculations on the quantization of the gravitational field, nor on the reality of the gravitino, the supersymmetric potential partner of the putative graviton.

USSR, Olga Oleinik and Nail Ibragimov

In my youth, few women, whatever their nationality, became researchers in mathematics. USSR made some exception with two remarkable women, Olga Ladyshenskaia and Olga Oleinik, both brilliant students of the soviet mathematician Petrovski, a professor in Moscow and essentially, my contemporary. The two Olgas worked, like me, on non-linear partial differential equations, with a particular interest in applications to physics. I knew them both and appreciated their works. Those of Ladyzenskaia were more related to mine, but I became a real friend of only Olga Oleinik. I met her for the first time in Paris; I appreciated her open and warm personality at once and thereafter, in all occasions when I could speak with her, even the last one. It was at the difficult time for mathematicians of the former Soviet Union when, already in bad health, she stayed briefly in Paris, invited by colleagues who wanted to help her. Before the perestroïka, in 1983, I visited her in Moscow. Thus I could, thanks to her, see some beautiful monuments of this big city. Olga had arranged, for me, a conference in the seminar of Vladimirov, a great mathematician interested in problems I had worked on. I had with him an interesting discussion.

During this stay in Moscow I had the pleasure of seeing Nail Ibragimov, a mathematician of great value and multiple interests: differential geometry, non-linear partial differential equations (he translated into Russian the Notes of Leray mimeographed in Princeton). I had known him in Paris, when he gave lectures in the Collège de France and we had gotten along well. Ibragimov knew Vladimirov. He came to listen to my conference in Moscow. Nail accompanied Olga and me in some visits of the city. He took us to visit the graveyard where lies the grave of the great mathematician-physicist Landau, known worldwide for his course on theoretical physics of several volumes written in collaboration with Lifshitz, that contains many original ideas. I don't remember if it is Nail who presented us the monument as "Landau-Lifshitz's grave"; it would not have pleased Lifshitz. In fact Lifshitz lived much longer, taking care of his health. In 1982 he was like me in China eating only provisions he had brought, wary of the Chinese food offered to us, though it was excellent. Ibragimov had come to Moscow, although he was now a professor in Ufa, the capital of the Bashkir Soviet Republic at the western boundary of the Ural Mountains. He invited me over for a few weeks. I accepted with pleasure, to get to know this region which was new for me. Nail was back there before my arrival. When my plane from Moscow arrived in Ufa, the stewardess told me a delegation with flowers was awaiting a personality at the foot of the plane. To my surprise, that personality was me. I never had had this honor (nor had it another time). Nail told me I was the first foreign professor to visit their university. I was very well received. Ufa was a big modern city that looked prosperous. Soviet domination seemed to manifest only as large colored posters representing the active forces in the country. Nail told me he was not from Bashkir, but from Tatar, a neighboring republic.

United States and Canada

In July 1989 I had the pleasure to be with Abe and Cece Taub in Boulder (Colorado) for the triennial congress, "General Relativity and Gravitation". Except for a pleasant chat with Cece, I do not keep any precise memory of Boulder. I remember Denver better, the capital city of Colorado and a big friendly town crossed by a busy main road traveled by a streetcar in free access, where I circulated alone for two or three days.

I spent November 1989 in Canada, invited by professor Ray Mac Lenaghan to Waterloo University. I had known and befriended him and his wife, a Belgian, during his visits to Paris. He was, like me, interested in properties of hyperbolic partial differential equations. We wrote, together, a Comptes Rendus Note on the diffusion kernel in Bondi coordinates. There was a polar cold but, after I bought a wonderful waterproof and windproof coat (I used it again in 2013 to visit the North Cape),

I went on pleasant excursions with the Mac Lennaghan couple, in particular to Toronto, a neighboring town.

The Montreal weather was mild in June where I walked on the occasion of a congress of a feminist association that had invited me there.

Work, Global Problems

In 1986, I worked again on the problem of the evolution in time of solutions of partial differential equations for fundamental interactions. I proved their global existence in a number of cases, for example in the neighborhood of Minkowski space-time, nil solution for harmonic maps (called non linear sigma models by physicists) in the case of 4 (or even 3 with Gu Chao Hao's collaboration) dimensional space-time. I extended the results, at the surprise of some physicists, to Anti de Sitter space-time, a strange Einsteinian model. Gu refused to cosign this extension, saying that his personal contribution was not enough to have his name there, modesty not very frequently encountered. I proved this global existence in the neighborhood of Minkowski space-time for Einstein equations in space-time dimension of at least six. I wrote, in collaboration with Mario Novello, under a form different from that of Helmut Friedrich, a conformal regular system for Einstein equations in dimension 4. With N. Noutchegueme I wrote a symmetric hyperbolic system for Einstein equations, with sources, rediscovered later independently by J. York. Together with Francis Cagnac and N. Noutchegueme, we proved an existence theorem for solutions of Einstein equations with data at the infinite past using the old idea of Christodoulou. With the decisive cooperation of F. Cagnac, I extended to a curved space-time a global existence theorem of solutions known for some non-linear equation on Minkowski space-time. Before seeing our work, Leray told me he doubted the truth of such a result.

Gravitation with Gauss-Bonnet Terms

At the end of the '80s, I thought that a classical supergravity could not have a mathematical formulation that was both rigorous and simple which could lead to physically interesting results. So I abandoned the work on supergravities and I looked at a theory becoming fashionable among some physicists: gravitation with Gauss-Bonnet terms. An English mathematical physicist, Lovelock, had found a Lagrangian reducing to that due to Hilbert and Einstein in space-time dimension four, but which differed in higher dimension, though it led to quasi-linear equations for the metric. Some works on this "gravitation with Gauss-Bonnet terms" opened problems without solving them. It was natural that the mathematical tactician that I am got interested in them. I proved in 1988 that the solutions of these equations do not satisfy Einstein's fundamental premise, causality. I presented this work in a seminar in Canberra, to the disappointment of our host, Robert Bartnik, who would have preferred that I speak of supergravity. But without wanting to compare myself to my great master Jean Leray, and unfortunately for me putting the phrase in the past tense, I would say as he did in an interview, "Only my current work interested me."

15

RETIREMENT 1990-2003

Dissolution of the Group

The ongoing restructuring of CNRS envisioned a decrease in the number of associated research groups. Our team in particular was up for renewal and it was clear to me that its multidisciplinary character would work against its continuation; my retirement, even if I asked for the three-year extension known as 'surnombre' in the French system, would remove any scruples the policy-makers might have otherwise had about its discontinuation. As for myself, I had received an invitation from the theoretical physics director, Philippe Tourrenc, to join his group, accompanied by some of the other members of the one I headed. Lichnerowicz had tried to encourage me to ask for the 'surnombre' but he could not convince me to reverse my decision. He regretted the dissolution of the group, knowing that he would not be readily accommodated by any other. In ours I had continued to let him feel like he was the boss, something no other director would have allowed.

The most qualified member of our group to succeed me, on the basis of his scientific merit and other qualities was Charles Michel Marle. After a remarkable thesis on relativistic kinetic theory under the direction of Lichnerowicz, this brilliant polytechnician redirected his efforts to the other pillar of his master's research, analytical mechanics and the algebraic structures associated to differential geometry. A group renewal proposal with Marle as Director was submitted to the CNRS *ad hoc* commission. The chairman wrote to Marle that his work was very interesting but advised him to ask for the creation of a group within the department of mathematics. This was effectively a somewhat ungracious dismissal of his application since it was obvious that, given the new CNRS policy of reducing the number of research groups, such an application had no chance of success. Hence our team was dissolved. Marle, for his part, was not overly chagrined at this outcome since he was already a professor at the university.

At about the same time though, Philippe Tourrenc reached the end of his tenure as Director of the theoretical physics group which was then housed in the Poincaré Institute, and Richard Kerner was asked to replace him. Richard accepted only after kindly informing me that he would only accept if I myself was included in this group. This was done, not only for me but also for some other members from the CNRS and even, thanks to Richard, for our administrative secretary, herself almost at age sixty, the retirement age. Mme Goudmand continued, even after retirement, to work, unpaid, for Lichnerowicz and myself, giving us inestimable services. In particular she insisted on compiling the reprints of more than 250 of my works. At my suggestion, with some regret, she kept only ten copies of each of them in colored folders on each of which she wrote the title and inventory number. She put these folders, grouped by two or three years, in appropriate boxes on which she wrote the relevant years and the numbers of the articles contained therein. She took care of having these thirty boxes brought to my house when I had to leave the large office that I had had in Jussieu. As I learned only recently she also handed over to the Académie's archives a folder containing all my publications up to 1995. Our conscientious and efficient documentalist Mme Serot Almeiras easily found work with another research group.

Retirement

I was sixty-six years old on December 29, 1989. Being a teacher, a mother of three children and not asking for the 'surnombre' extension, I finally retired on October 1, 1990, without particular joy but without regret. Lichnerowicz, supported by my friends and former students, convinced me to accept the holding of an international colloquium on this occasion. He chaired the organizing committee composed of Lise Lamoureux-Brousse, Richard Kerner and Moshe Flato, whose relationship with the banking community allowed him to bring in useful financial support. After consultation with Lichnerowicz my best collaborators and friends could thus be invited to this colloquium at the Collège de France in 1992 and to the following dinner buffet in the beautiful room on top of the Zamansky tower. The conferences, assembled in written form by Flato, Kerner and Lichnerowicz, were published by Kluwer Press under the title "Physics on Manifolds".

As a member of the Académie I was automatically emeritus professor, an honor which gives the right to retain one's working means, in particular an office. For some time I received from the administration, each school year, a letter informing me that my title of permanent emeritus professor was being renewed for the year! However the direction of my UER (Unité Enseignement Recherche) had exchanged the corridor where my office and the secretariat-library of our research group were located, with that of our neighbor the Numerical Analysis laboratory for some advantage that, as I was later told, they never got. I was not kept informed about these negotiations.

The researchers of CNRS had to vacate the place. Marle obtained an office among mathematicians and a room to use to make a small library with the mathematical documents removed from ours. A small party was organized in this room to celebrate the opening. Lichnerowicz accepted without protest the unmerited thanks for this gift. I knew him too well to blame him. However I am grateful to Akbar Zadeh, one of our best researchers and a specialist in the study of Finsler spaces, who said to me, "But it was you who created and filled this library".

The then-Director of the Numerical Analysis Laboratory, Philippe Ciarlet, my colleague in the Académie, came to see me and said, "As long as I am Director you will keep your office". I was thus able to keep my large office in tower 66 for some years. When Ciarlet was replaced at the direction of Numerical Analysis by Olivier Pironneau, however, some pressure by intermediaries was exerted on me to accept the splitting in two of this office in order to accommodate a member of Numerical Analysis which was now cramped. I resisted for a while. However the occupancy plan of the Poincaré Institute was redesigned and the theoretical physics group of Paris VI University was relocated to tower 22 of the Jussieu campus. Pironneau went to see Richard Kerner asking him to find a place for me. I then left my big office in tower 66 to mechanicians, who never really considered me as one of them, for a smaller one, but among physicists. Several of them were old acquaintances and friends: Richard Kerner of course, but also among others whose names do not immediately come to my mind, Phillippe Tourrenc, Bartholome Coll, Bernard Linet, Michel Chevreton and also Christian Bordé whom I now have the pleasure of meeting in the Académie. I retain pleasant memories of my tenure in tower 22, of the theoretical physics seminar sessions, of the recommendations of Philippe Tourrenc concerning the replacement of the program "Mathor" by "Scientific Work Place" and the use of the shared computer and, above all, the frequent interesting and friendly discussions with Richard Kerner and his visitors. Unfortunately, the time finally came for the removal of asbestos from tower 22 and the necessity to leave. I then made my farewell to the Jussieu campus and the working facilities it had offered me. I must say that the Pierre and Marie Curie University continues to forward the emails which are sent to me there to my new address at the Institut des Hautes Études Scientifiques (IHES) in Bures-sur-Yvette, and I am grateful for that.

Private Life

Our life in Antony continued almost without change for Gustave and myself, except for the birth of our grandchildren that pleased us both very much. First Mahé in 1990, always smiling and affectionate, both as a baby and then as a little girl. I often went to pick her up at the day nursery on Friday and bring her to Antony; her parents would then come on Sunday to take her back home to Paris. In good weather Gustave and I brought Mahé to the park in her stroller. At home I played with her often and still see myself, when she was a little older, playing lotto with her, she lifting the number and saying with a charming smile "c'est qui?" ("it's who?"). I remember my pleasure when she would say to her mother who had come to fetch her, "Je ne veux pas rentrer à la maison" ("I don't want to go back home").

Fortunately her mother, an intelligent woman full of common sense, would not look offended. Mathurin was born two years later. We then had both of our grandchildren for the weekend fairly often, always with pleasure. To my regret, Daniel's last son, Enzo, was born in Bordeaux, and the three sons of Geneviève, Axel, Timothée and Andreas were born in Lyon. I therefore benefited less from their childhood.

Work and Travel

Being still physically and intellectually sound, I have worked well during my retirement, and I have had the pleasure to go to various countries where I was invited to work in collaboration or simply to present seminars of general interest on my works. In June of 1990 I was happy to see Montreal again, on the occasion of a congress organized by a feminist association. I will speak further in a later chapter of my travels to distant lands.

In 1990 I traveled a lot in Europe. I spent February in Catania, invited by Marcello Anile, who was afraid that my retirement would deprive me of the material help necessary for the writing of my work. In fact computers and software now render this help superfluous but I nevertheless rejoiced in the opportunity to discuss problems of common interest concerning relativistic fluids, and potentially to collaborate with Marcello, whom I much appreciated scientifically as well as on a personal level. However when I came to Catania Marcello had changed his center of interest and so we did not, after all, work in collaboration. We nevertheless spent some good time together and I enjoyed this stay in Sicily. I returned to Italy in the spring, this time to Varenna, magnificently situated near lake Como, to participate in a congress on partial differential equations organized by Jalal Shatah and Sergiu Klainerman. In September I again had the occasion to admire the beauty of our Latin sister during a congress on "Waves and Stability in Continuous Media" organized in Capri by Giuseppe Marmo, a professor in Naples.

Artur Alves, a student of Ribeiro Gomes in Coimbra who had, under my direction, written a thesis on relativistic kinetic theory, came back to spend some time in Paris and worked there with Henri Cabannes on discrete kinetic theory. Together they organized a colloquium on this theory in Figuera da Fos, a touristic town by the sea, not far from Coimbra. Guy Pichon had found an original way to formulate the theory on a curved space-time and I had seen how to obtain an existence theorem therefrom. We were both invited to the colloquium and had the pleasure there to awaken the interest of a world-renowned Italian specialist on kinetic theory, Carlo Cercignani. Pichon and I, in collaboration, then wrote an article for the proceedings of the colloquium. When leaving Figuera da Fos I made, together with Pichon, a short visit to Coimbra. Walking with him along the Mondego river I told him of my nostalgia to see this landscape for the last time. He was surprised because he found nothing very remarkable in it.

In 1991 I attended a symposium in Elba organized by my friend Giorgio Ferrarese. His wife Liliana had befriended a couple who had, in this pretty island facing Livorno, a nice hotel equipped for small conferences. I went there several times, always with pleasure and interest.

I went to Spain several times, invited for conferences at the autonomous university in Madrid by Carlos Moreno, a former student of Lichnerowicz, who was a part-time member of our research group. Though we never worked in collaboration we did discuss a variety of subjects and I became real friends with him and his charming wife. I remember these discussions with pleasure as well as my stays in Madrid and my visits to its wonderful museums. I took advantage of my invitations to Madrid to go visit several of the beautiful cities in Spain, Burgos and Seville and others. The most striking memory I keep from these visits is the impressively beautiful cathedral of Cordoba, a former mosque. I remain grateful to Carlos for these invitations.

Spiros Cotsakis and the Samos Meetings

Spiros Cotsakis, born in Athens, is forty years my junior. He is a direct and warm man and a great organizer and we easily became friends. After writing a thesis on mathematical problems in General Relativity and cosmology in England, he returned to his native country and became professor at the University of the Aegean in Karlovassy, a town on the magnificent island of Samos. Nevertheless Spiros spent a lot of time in Athens where his wife was working and his family actually lived. In 1994 Spiros organized the first "Samos meeting" in Karlovassy and invited me to come, perhaps on the advice of another Greek, my friend Demetrios Christodoulou. I was happy to accept. Greece is a beautiful country, the Greek islands all different and magnificent. Samos is covered by lush vegetation, a beautiful green in the middle of the very blue Mediterranean. Pythagoras had lived there. Spiros and I got along well, earning me invitations to the second and third Samos meetings and even to a congress organized on the other Greek island, Lesbos, by a former student of Spiros, Alessandro Mirizzi. In Lesbos I was happy to get better acquainted with the very amiable second wife of Demetrios who is also Greek. She did not attend the conferences but walked in the neighboring countryside and brought me back a big bunch of oregano which long perfumed my kitchen. A friendly and intellectually stimulating atmosphere always reigned in these small meetings held in such magnificent landscapes.

Global Problems, Collaboration with Vince Moncrief

Vincent Moncrief, called Vince, is an exceptional mathematical physicist. It is a pleasure to work with him because he combines physical intuition and mathematical rigour with new ideas that lead to solution of real problems of great generality. I was lucky that Vince, a confirmed bachelor, likes Paris and willingly stays there. Thus we began our work together. He had obtained results on the local in time construction of Einsteinian space-times possessing some symmetry by using the Hamiltonian method, dear to his heart. I showed that one obtains easily the same result with a more direct method which is better suited to the search for global existence in time. We then began to work together, but stumbled upon a difficulty which prevented us from succeeding and we temporarily abandoned the subject. A few months later an original idea of Vince allowed us to resume the project. Vince passed on to me his large sheets of lined yellow 'legal' paper, without which he could not work, filled with calculations which I resumed, completed where necessary, and began writing a synthesis of this first global theorem with a compact initial manifold. We worked together fruitfully, in perfect harmony, when we had the opportunity — in Paris, in Samos and also in New Haven. Indeed Vince, a professor at Yale University, had obtained a collaborative NSF-CNRS grant that funded our joint research. I have good memories of my stays in New Haven, housed in a pleasant residence for researchers in theology. Vince took me on a Sunday for a tour of the neighboring countryside and told me, apparently with surprise since he is not normally a sightseer, that it had been very pleasant. I also sometimes walked alone in the town. Once near the town center, being a little away from the more populated areas, the atmosphere seemed to me worrisome. Encountering a young black man casually eating a sandwich, I asked him if I was going the right way. He answered kindly but added, "But don't stay around here very long, it's dangerous". I followed his advice. It seems though that New Haven now is as safe as the other cities of the USA.

Chapel Hill, Collaboration with Jimmy York

I first met James York, known always as Jimmy, in Princeton in 1973. Subsequently he obtained a professorship at the University of North Carolina at Chapel Hill, replacing Bryce DeWitt who had moved to the University of Texas. In 1979 I proposed to him that we collaborate on the chapter on the Cauchy problem in General Relativity that Alan Held had asked me to write for the book he was editing for the occasion of the centenary of Einstein's birth. Jimmy, as a collaborator, was efficient and pleasant to work with. I was a little upset though when he wrote shortly thereafter a chapter on the same subject for a book edited by Larry Smarr, a specialist on numerical analysis, without consulting me. This article is much more often quoted than ours because numerical analysis has taken on, in relativity as elsewhere, a development of added importance with the considerable technical advance of computers. Even so I never complained to Jimmy about this.

In 1993, my son Daniel and his wife were invited for a two or three years stay at the university in Durham, the city in North Carolina not far from Chapel Hill. Wishing to make some visits in the neighborhood of my son and grandchildren, I wrote to Jimmy to suggest that we apply for a CNRS-NSF collaborative research grant like the one that had allowed my joint work with Vince Moncrief. Jimmy replied that he did not wish to increase the weight of his administrative duties but that he would very gladly invite me with the funds available to him thanks to a bequest to the university in Chapel Hill by Agnew Bahnson. In the subsequent years I made a few pleasant visits to North Carolina. During the week I worked with Jimmy, housed in a small apartment of a house located in a leafy suburb. The friendly owner, Hope Rice, was the widow of a professor. The apartment, arranged years before to accommodate her mother in law, was very convenient. I had rented a car and was spending the weekends with my son Daniel and his family, enjoying the delightful presence of my grandchildren, two charming toddlers, sometimes going to fetch them at kindergarten.

With Jimmy and his students Andrew Abraham and Arlen Anderson we extended to general coordinates the result I had obtained with Tommaso Ruggeri. After the departure of Andrew we proved rigorously, sometimes with Arlen or just with Jimmy, various general theorems useful in the solution of Einstein equations that have since become increasingly important with the revolutionary progress of cosmic observations made possible with satellites and powerful telescopes.

I had not really befriended Jimmy's first wife whom I had not seen much. We had conversed a little only when she accompanied Jimmy at a congress in Samos. She then told me with astonishment, "I thought we would have nothing to say but we can speak together about our grandchildren". I was surprised, however, when the couple separated because Jimmy seemed to be, as a common friend had said, a "family man" very attached to his two children. His second wife, Sarah, has always been very kind to me, in particular when Jimmy invited me, with NSF support, for a collaboration at Cornell where he worked for some time after his divorce and his departure from Chapel Hill.

Completion of a Work with V. Moncrief, Charpak in Cargèse

I have already mentioned that Piotr Chruściel and Helmut Friedrich organized in 2002 a summer school subtitled "Fifty Years of the Cauchy Problem". For the book containing the texts of the lectures I completed the work I had begun to generalize the global existence result obtained with Moncrief. Vince had written the introduction, but when I had proven the desired result, he refused to cosign the chapter containing my result, saying that his contribution did not deserve any credit for it and that in any case, thanks to another result obtained in collaboration with Lars Andersson, his name would already appear in this book dedicated to me.

That same year my friend Georges Charpak spent the holidays in his house in Cargèse together with his wife, a very pleasant woman whose generous and unconventional character accorded very well with those aspects of that of Georges. The lunch they offered me on their terrace with a splendid view of the Mediterranean is of my best memories.

16

A l'I.H.E.S 2003-?

The IHÉS

The IHÉS (Institut des Hautes Etudes Scientifiques) is a private organization dedicated to advanced research in mathematics, theoretical physics and any other science linked with it. It was founded in 1958 on the initiative of Léon Motchane. After a successful career in industry, Motchane, always passionate about mathematics, had written a thesis initially under the direction of Arnaud Denjoy, a mathematician with varied interests and a very open mind, but who was already old. He soon retired and asked his former student, Gustave Choquet, to replace him in directing Motchane. Gustave accepted so as not to annoy Denjoy, but he was a man little used to concessions and passionate in mathematics. He suffered to see their rigour somewhat mistreated in Motchane's enthusiastic researches. The persevering work of Motchane resulted, however, in a doctoral thesis of Paris University. The industrial career of Motchane had left him a fairly large capital; he had, also, relatives well-provided. One of them offered, to "Collège de France", half of the funds necessary to create a new Chair, probably hoping that it would be attributed to Motchane, but this noble, state-dependent institution rejected that offer. This was fortunate for French science since Motchane then envisaged creating his own institution in the image of the famous American Institute for Advanced Study (IAS), which operates on private funds. Encouraged by the director of the IAS, Robert Oppenheimer, and Cécile DeWitt, who at the same time put forward her school of theoretical physics in Les Houches, Léon Motchane realized his ambition to be a great servant of mathematics. The "Institut des Hautes Etudes Scientifiques" has developed, since its creation, according to the vision of its founder, Motchane, financially helped by other industrials of his family or their friends. He found a small castle in a great park in Bures sur Yvette, and arranged in it a conference room, library, some common rooms and offices to make it a privileged place of work, meetings and collaborations. From the beginning, there reigned an atmosphere of free discussion among scientists, whatever their specialty or origin was. The statutes of IHÉS are:

"Art. 1: The Foundation called "Institut des Hautes Études Scientifiques" aims to perform and to promote theoretical scientific research in the following domains: Mathematics, Theoretical Physics, Methodology of Human sciences, and any theoretical discipline linked with them."

"Art. 2: The action of the Foundation is mainly to place, at the disposal of professors and researchers of the Institute, permanents or visitors, resources allowing them to realize disinterested research: a center including offices, rooms for work and meetings together with a library, a cafeteria and a residence to house temporary researchers."

Motchane had the intelligence to ask to come, as mathematicians first, two very different scientists, both of exceptional and incontestable value — Jean Dieudonné and Alexandre Grothendieck. Soon afterwards he added René Tom, also remarkable, then the theoretical physicist Louis Michel, the first French to have escaped the backward grip of Louis De Broglie, and the incomparable David Ruelle, famous for his fundamental work on chaos theory. The choice of these first permanent members accredited, at once, the IHÉS in the scientific community. The scientific directors replacing Motchane, the Hollandese, Kuiper, then the French, Marcel Berger, have been able to preserve the IHÉS' excellence and enlarge its premises, thanks to generous donors. The excellent differential geometer, Jean Pierre Bourguignon, who succeeded them in 1994, not only preserved this excellence but also worked, with much energy and success, to further the development and international fame of IHÉS. Thanks to other donors, among them the great mathematician S. S. Chern, Bourguignon had arranged the construction, arranged in the park, of an amphitheater for important conferences which was preceded, upon the advice of Cecile DeWitt, by a large meeting room where cocktails are served on relevant occasions. There are, at IHÉS, still very few permanents, all members serving as professors in "Collège de France" of the French scientific elite. Now there are also, for brief or long stays, many visitors, originating from all places in the world, whose ages ranged from young ones in the beginning of their career to scientists already famous.

Thibault Damour became, in 1989, one of the few permanent professors at IHÉS. He is an exceptional theoretical physicist. His immense knowledge in both physics and mathematics always surprises me. We owe him fundamental results in very different directions, from the rather abstract string theory to the solution of very concrete complex problems which have since received remarkable observational confirmation: a few years ago, the slowing of a pulsar by emitting gravitational waves; more recently, the spectacular observation of such waves, gravitational field oscillations resulting from the collision of two black holes. The signal, observed after almost fifty years of efforts, agrees with the prediction, through reasoning and difficult calculations, by Thibault Damour and his collaborators, in particular Donato Bini and Alessandro Nagar.

Eighty Years

On December 29, 2003, I reached 80 years of age, old of course, but still sharp. With the agreement of Jean Pierre Bourguignon, Thibault Damour organized, in IHÉS, a colloquium to celebrate my birthday. The then-President of the Academie was Emile-Etienne Baulieu, a doctor and biochemist, famous, in particular for his discovery of RU 486, the morning-after contraceptive, which changed life for many women. Thibault thought Baulieu would not come for me, a mathematical physicist and contacted the Vice-President, Edouard Brézin, an author of important works in theoretical physics, a subject closer to mine than biological discoveries. In fact, as soon as he was candidate to the Academie, I had sympathized with E. E. Baulieu and was always happy after his election to meet this brilliant, always friendly colleague, who was a feminist, full of spirit. Therefore, I was not surprised that Baulieu bothered to come to preside over the first session of my birthday colloquium. Brézin gave me also the pleasure of being there, as well as did my friend, Remo Ruffini, creator of the "Marcel Grossmann meetings" who now had more than a thousand participants assembling under his leadership. On the first day of the colloquium, Remo gave me the beautiful silver object representing a vortex around a black hole, a prize awarded to a fortunate elected relativist at each Marcel Grossman symposium. I had been chosen for this prize at the last of these meetings, in Rio de Janeiro, but I had not been able to go to

Brazil to receive it because I was participating in a program on general relativity in Santa Barbara, California then.

Hospitality at IHÉS

On the occasion of my birthday colloquium I much appreciated the nice cooperative atmosphere reigning at IHÉS, in welcoming premises favorable to work. Asbestos removal from Jussieu towers had made me a physico-mathematician without an office nor Internet means. Chatting with Jean Pierre Bourguignon, I asked him if I could come occasionally to IHÉS to use the collective computer and printer. He answered yes but added that I should have there a permanent base. Thibault found a solution by asking Pierre Cartier, retired from CNRS, to share, with me, his office. He probably could not refuse, but he accepted with good humor. I had known Pierre for more than forty years and always had excellent relations with him. So I had the use of a coat rack and an armchair as well as a piece of a table in the office, full of books and documents, which had been occupied by Pierre for many years. It was perfect for me. The hospitality of IHÉS has changed my life in these last years. Thanks to it, I remained in contact with the science in progress and its propagators and perpetrators who were or became my friends. Especially pleasant for me are my excellent weekly lunches in its cafeteria with meals cooked by Patrick, and always kindly being served first by Dominique, and by other kind ladies when she retired. I used to sit at a middle-sized round table between Thibault Damour and another friendly scientist. It was often Laurent Laforgue, a physicist and a mathematician with the age of one or the other of my daughters. Interesting and varied exchanges with them and other guests stimulate my aging mind. Thibault continues to enlighten me with a commendable lot of good will on the rapid, often surprising development of our shared mistress, General Relativity. The tea, served in the common room by nice women, is another occasion where interesting discussions were had, also with temporary visitors.

The informatic service of IHÉS, François Bachelier and Karim Ben Abdallah, gave with great kindness, precious help to the neophyte I was in the use of computer equipment and programs necessary today to researchers, even theorists. The secretaries, especially Elizabeth Jasserand and Marie Claude Vergne, have always shown me a comforting sympathy. I pass the open door of Marie Claude's office to reach mine, and she became a real friend. When our schedules permit, we chat while she drives me to the RER station.

Work

I continued my work as a researcher after 2003. I had the good luck to be able, in collaboration with some younger researchers, to obtain progress in the solution of some open problems. With Jim Isenberg and Daniel Pollack we solved the Einstein constraints with a scalar field source; with my long-admired friend, Helmut Friedrich, I had the pleasure of writing a small article on a particular case of the difficult problem of the n bodies motion in General Relativity. I got interested in the behaviour of Einsteinian space-times in the neighborhood of a singularity, for instance, the Big Bang. Important works, though difficult to formulate in precise mathematical terms, concerned the study of the case of chaotic behaviour. In collaboration with J. Isenberg and V. Moncrief, I became interested in the opposite case where the space-time tends towards a solution of equations deduced from the Einstein equations by suppressing spatial derivatives. We then obtained results after proving a general theorem we called the Fuchs theorem; it appears in Appendix V of my 2009 book. In IHÉS, I spent much of my leisure time writing that book "General Relativity and the Einstein Equations" which brings together a number of my works on mathematical problems posed by the Einsteinian theory of gravitation and its sources. I like to write, that is, to attain a precise and clear expression of a thought. It is more possible in mathematics than in any other domain, but personally I find it more satisfying if I can imagine that it portrays part of this strange reality we live in.

My book nearly finished, I offered it for publication to Elsevier. I received, in response, an acceptance contract proposal that I hesitated to sign, because it imposed on me a material finition work which I found excessive. In the meantime, I had the chance to meet John Ball, who had come to listen to my Emmy Noether conference at the international mathematical congress at Madrid in 2006. I told him of my perplexity. John was a member of the editorial board of "Oxford Mathematical Monographs". Soon afterwards I received a letter from a charming young lady, Alison (I don't know her family name; nowadays everybody uses first names) telling me that if I had a book on General Relativity, Oxford University Press would be happy to edit it, that is, publish it after professional work for formatting the text. I accepted with pleasure. Unfortunately, Alison resigned from Oxford University Press. My text has been edited and published in 2009, but the presentation may be less enjoyable than it could be. My book had flattering analyses in specialized journals but was not a great success in bookshops. The managers of Oxford University Press probably changed, because I later received a very nice email from a gentleman named Keith Mansfield, telling me the presentation of my book is not very attractive, but that its volume makes a paper edition impossible. He suggested I write a simplified version which could be more attractive. He added kindly that he had heard me give a lecture which he had much appreciated. I accepted, quite willingly, his proposal, which gave me interesting work that was less difficult than research of original results in a domain where competition is now very strong. I told my friends that my intellectual capacities were less great than before and that I would put, in this new book, only what I still understood. I took pleasure in writing the book, "Introduction to General Relativity, Black Holes and Cosmology" (I had wanted to put "basic" and not "introduction" but the commercial division of O.U.P. insisted on the word "introduction"). I suppressed the heavy mathematical proofs, added physical comments, and quoted the new theoretical, observational or experimental results, now thriving. Members of the Oxford University Press did all they could, under the encouraging direction of Keith Mansfield, to make my book attractive. For the cover, Keith accepted a brightly-colored abstract painting entitled "cosmos birth", the work of my daughter Geneviève, which made the book more pleasant to see. I thank Keith for the fruitful interest he brought to the book he had advised me to write.

Centenary of General Relativity and Observation of Gravitational Waves

My last scientific writing was a chapter on the beginning of the Cauchy problem that had been requested by the active and nice young relativist Lydia Bieri, in the book she edited with S. S. Yau to celebrate the centenary of General Relativity. Recently, I wrote a brief report on my meeting at the I.A.S. with Einstein, for a colloquium organized in Berlin on the occasion of this centenary by the director of the group of gravitational physics at the Max Plank Institute, Herman Nicolai, a specialist world-renowned for his works on gravity and quantum theories, some in collaboration with Thibault Damour. My report was published in an Internet annex of the journal "Gravitation and General Relativity", directed by the specialist of the history of General Relativity history, Clifford Will. Here, I thank my friends Lydia and Clifford.

The centenary of General Relativity coincided up to a few months with the observation attempted, for half a century, of gravitational waves — that is, rapid oscillations of the gravitational field propagating with the speed of light — as heuristically foreseen by Einstein and mathematically proven in my thesis to be implied by his equations. Gravitational waves, contrary to electromagnetic ones, are not stopped by any obstacle. They should be able to give us new information on the universe we live in and even on the internal structure of stars. I am now too old for original scientific work and even, to follow the details of work made by others; however, the turmoil caused by the recent announcement of the observation of gravitational waves after considerable work, did not escape my notice. I have been led to accept a few interviews, though the future is for me, like for everybody else, impossible to foresee.

17

FAR AWAY TRAVELS

During my career as professor-researcher, I took a lot of interesting trips, often financed by those who invited me, sometimes by the Ministry of Foreign Affairs in developing countries. Grants were often given upon request by the scientists wishing my visit, sometimes through the International Relations Committee of the Académie whose then-president, the physicist Guinier, was well-disposed towards me. In this chapter I evoke some memories both of my rewarding travels in distant countries and enjoyable interludes.

India

My mother appreciated, after her retirement and before the deterioration of her health, the traveling she did with her son — a young mathematician precociously famous — in exotic countries where he was invited. The first trip had been to India in 1958, at the Tata Institute in Bombay where worked a great mathematician, Narasimhan. My mother was happy to go to India, having always been very interested in its civilization. I was teaching in Marseille then, but living in Les Lecques. My brother had traveled by plane, a mode of transportation still unfamiliar at the time. Fearing for his mother's heart, François had organized for her a boat trip, leaving from Marseille. My mother did not wish to take this opportunity to see me in Les Lecques, but wanted me to come to Marseille to meet her at the station and accompany her to the harbor the following day. I did as she wished, as I always did. I slept the day before her departure in the same hotel as her, near the train station, where she had booked two rooms for us. I took her to embark on her trip the next day. She kept a good memory of her sea trip. She also appreciated the beauty of Hindu monuments, the kindness of the professors who had invited my brother, and the magnificent hotel Taj Mahal where they were housed, but she was shocked by the misery that reigned in the streets of Bombay. At about the same time, Marcel Brelot was also invited at the Tata Institute and went to Bombay with his wife, who told me later that she never went out of the hotel throughout her stay, so depressing was the spectacle of the street!

These mixed feelings that my mother had kept from her trip to India did not prevent me from wishing to accept the invitation I received to attend the international congress organized in Calcutta in 1975, to honor the memory of Professor Bose, an Indian physicist famous for the statistics called "Bose-Einstein". The Ministry of Foreign Affairs offered to subsidize the travel costs of three French relativists. Lichnerowicz, fearing for his health, did not wish to go to India. Crumeyrolle and another former student of Lichnerowicz, Pigeaut, a professor in Dijon, did. We met in Calcutta, with relativists of other nationalities. I forgot their names, except for the Brazilian, Mario Novello, a young and dynamic cosmologist with whom I sympathized. He became my friend and we met several times thereafter in Rio or in Paris.

The aircraft rates in 1975 were not what they are today. It was not more expensive, and sometimes necessary, to make several stops on the way. The ministry did not object. On the way to Calcutta, I stopped first in Tehran. I did not go to Isfahan, thinking to visit this famous place later. In fact I never had the opportunity, which was too bad. The Tehran monuments do not compare to those of Isfahan, but I am glad for this brief overview of the Iran of 1975, a city of contrasts. I have walked along a large modern street lined with tall buildings, many of them banks, and a parallel street with small shops that were typically oriental. I entered, for lunch, in a coffee shop of modern aspect where I found myself to be the only woman. However, the server came to me with a pleasant smile and handed me a menu in English for me to make my choice. I lunched soberly while none of the others present seemed to remark on my presence. After Tehran I made a brief stop in Karachi, which does not leave me a vivid memory, renouncing Lahore as I did Isfahan. Before joining Calcutta, I stopped for two days in Delhi, then in Agra, and at last Benares, now called Varanasi. I slept in local hotels that were inexpensive, though comfortable. Delhi is an interesting city, rich in ancient monuments. In Agra, the visit of the Taj Mahal, in a garden where women with varied colored saris walked slowly, is an unforgettable memory. (I was disappointed, years later, when I saw again the Taj Mahal, overrun by tourists.) From Agra, I wanted to visit Fatehpur Sikri. On the advice of the innkeeper, I went to take a bus there. In that bus, I met an Indian couple who told me I should have taken the tourist bus where a guide would have guided me across the splendid monuments of red sandstone in this imperial city. They were there themselves to visit the place and offered to be my guides. The husband was an engineer and spoke perfect English. I spent an excellent afternoon. When we parted, we exchanged names and addresses. I hoped to see them in France, but it did not happen.

Varanasi was very picturesque: I crossed, then followed a joyous crowd up to the Ganges, accompanying a stretcher carrying a dead person who was going to be cremated. This display did not really move me; I felt too foreign to it.

In Calcutta, we were housed in a building in a large flowering park. The most vivid memory I keep of Calcutta is, when we left this refuge, the view of the human tide that flowed in the streets of the city.

On the way back, I stopped in Nepal, at Kathmandu. It was interesting to see these constructions of another civilization, and the young girl, the future holder of a title in her religion, I don't know which, who saluted from the window of a house from which she was not allowed out. I took a bus that led us to the Chinese border. The most vivid memory I keep is of peasants washing their vegetables in the dirty stream in front of their house. I had problems leaving Nepal in spite of the Air France ticket I had. The desk clerk that I addressed, upon arrival at the airport to confirm my ticket, told me the plane was full for a month! Walking in the town, I saw a Thai travel agency. I went in to look for another flight. The employee, a charming young Thai, took my Air France ticket, made a passage in the backroom and brought it back with a large smile, confirmed for the expected date. I was told that the Air France clerk expected a baksheesh for validating my ticket. I am grateful to the young Thai lady for her smiling and selfless act.

Though I did not make a true personal friend in India, I returned there several times because that country has a long tradition of remarkable mathematical physicists interested in General Relativity. I was invited to give a conference in several congresses on this subject. Invited in 1984 to celebrate the centenary of Sir

Arthur Eddington's birth at Nagpur, I asked for a grant for travel costs from the Ministry of Foreign Affairs, which was given to me under the condition that I also visit other universities. The Ministry would cover not only my transports, but also my living expenses in these other universities if I received invitations from them. It was easy and I was glad to visit the south of India which I did not yet know. I proposed my visit to the mathematician Narasimhan at the Tata Institute in Bombay, and to the physicist Vishveshvara in Bangalore, both scientists known worldwide. They warmly answered and sent me invitations. I do not remember how I went to Nagpur and I have only a vague, but pleasant, memory of the city and the congress. I remember that after Nagpur, I had gone to the fairly close city, Hyderabad, in the company of the friendly Crumeyrolle, who had been leaving from there to France. He told me he was surprised that I undertake my journey alone; his wife would not have done it. It was, however, easy to take the plane for Bombay, where Narasimhan had reserved, for me, a room in the luxurious hotel Taj Mahal. I think that its price now is no longer accessible to scholars. Narasimhan offered me the best meal in a restaurant I had in India. I had not written to the relativist Narlikar, whom I had not met before, but he came to see me and we discussed cosmology with interest. Like my friend Irving Segal, Narlikar does not believe in the standard cosmological model, making him now a bit of a pariah among relativists. As for me, I think that our strange universe still reserves, for us, many surprises. The existence of alternative theories is stimulating, thus healthy, for a necessarily incomplete science like cosmology.

In Bangalore, I was hosted by a local institute, where I lived the way some Indians of the time did. It was an interesting experience, though the shower with two heads, one for almost boiling water and one for really cold water, was not ideal comfort.

Also the vegetarian food, everlasting rice with a horribly spiced sauce, had no chance of making me fat. To go to town from this housing, you had to cross a smelly stream on a board without a parapet. Fortunately I was still fairly young and I did not lose my balance. I am happy to have been able to know this region. Contrary to my stay in Bombay which did not have for me much touristic interest, my visit to Bangalore had made me discover an India very different from that which I had seen during my first stay. The richness of the decoration of the temples of the south is much more abundant than in the north. In fact, while appreciating it for its value, I was less sensitive to the beauty of the temples of south India. However, I remember the admiration I felt for the two touristic sites I had visited in the neighborhood of Bangalore with a local coach.

My next visit to India, in 1986, introduced me to yet another region. The physicist D. B. Pandey, a specialist of General Relativity I had met at an earlier congress, had arranged my visit to Gorakhpur where he was a professor. I gave there a lecture on a remarkable recent work of Christodoulou which had attracted my interest. It concerned the weak global solutions with spherical symmetry of Einstein equations. I think these difficult mathematics disappointed listeners, though they had the politeness not to show it. Gorakhpur is a city of Uttar Pradesh, a state of North India, near the Nepal border and the birthplace of Buddha. My most vivid memory of this stay is my excursion to visit the place that commemorates this important event in the history of humanity. There was no spectacular monument which I remember, but an atmosphere imbued with spirituality.

After leaving Gorakhpur, I stopped as a tourist in the capital city of Uttar Pradesh, Lucknow, which leaves me only a vague memory, pleasant especially with my satisfaction at having managed to arrange, by myself, this stop with a comfortable hotel for a very reasonable price. Before taking my return plane from Delhi, I went to visit the beautiful city of Jaipur.

Another stay in India occurred in 1989, when I was invited by a former member of our team of relativists, Husein, who had become director of the physics department in the Aligarh University. It was interesting to see this Indian city not transformed by tourism, and the attentions of Husein, who lived in Western style, made this visit very pleasant. Husein came, in turn, to visit me in Paris. He had written to me asking to postpone this visit for health reasons, then written again, saying his doctor had advised him to perform it as scheduled. He came in March 1990, accompanied by his daughter, a courageous and friendly woman who was going back to the States where she is a doctor. They were happy to visit the new buildings of the Paris VI University and admire Paris from the top of the Zamansky tower. Husein accompanied, to the airport, his daughter, who was flying at dawn to the States before he himself went back to India. It was a farewell. They knew it; probably because Husein was diagnosed with pancreatic cancer, he died a few months later. I think back to their stay with great sympathy.

My last trip to India was in 1991 for a congress on General Relativity in memory of Professor Vaidya, in Ahmedabad. I found there the modern atmosphere of big cities, crowded and with intense circulation of cars. Thibault Damour was also at that congress, I remember, better than what I saw, the pleasure I had to visit Ahmedabad in his company and my admiration for his ability to find his way. Before leaving India, starting from Delhi, I wanted to see again Agra and Varanasi. These cities and their inhabitants were transformed since 1975. These changes are not an advantage for tourists; I hope they are for the inhabitants. India is a great country; Indians have brought much to humanity in the past, and I wish them a future as happy as it is possible in this world.

Togo

I spoke before of my stays for work or tourism in the Maghreb countries. My stays more to the south of Africa come from my scientific relations, except that I did, in Togo in April 1977, make acquaintance with the eldest of my grandchildren, Raphaël, and keep his mother company. Her companion, a mathematician teaching at Lomé University, was on leave for a few days after I gave a lecture at the department where he taught. The memories I keep of my fairly short stay in Lomé are neither touristic nor scientific but those of the daily life, pleasant but ordinary, with a very young baby. It is different from my subsequent stays in Cameroon or South Africa.

Cameroon

His thesis obtained, Francis Cagnac, who was a priest, wished to combine his missionary vocation with his mathematical gifts. With the approval of his bishop, he solicited and obtained a professorship at Yaoundé University, in Cameroon, in the framework of cooperation which existed then between France and that country. "Father Cagnac", as he was called by his Cameroonian students, has dedicated himself selflessly and tirelessly to the teaching of mathematics to Cameroonian students, and directing in research those who had the ability. Cagnac succeeded in creating, in Yaoundé, a valuable school of mathematics, a feat I think exceptional in that part of the world. Cagnac, however, felt a bit alone in facing his job as Research Director. I believe he has been glad to find in me a travel lover. Facilitating his relations with the evolving mathematical world, I was able to support him in his task, and happy to do it.

Thanks to the cooperation funds, I made stays of about a month in Yaoundé, first in 1983, then in 1986. The CFA Franc existed then and France was generous to cooperants. I was housed at the Sofitel and I had a car with a driver for my trips between the hotel and university. The atmosphere in town was peaceful. I walked alone, amused by the market's animation. Cagnac had, I believe like other cooperants, a comfortable salary. However, he lived very simply and had only a "quatre chevaux" Renault to drive. This did not prevent him from taking me on a few interesting trips in the neighborhood, stopping sometimes for a refreshment at the place of some clergyman he knew.

Cagnac directed several doctoral theses in domains of our common interest. During my stays in Yaoundé, I could hold discussions with Cagnac's students and sometimes help them, or at least encourage them. The first was Norbert Noutchegueme, who did a thesis on the difficult problem of the solution of Einstein equations with data at past infinity. I had begun this study in Torino in 1978, in collaboration with Francaviglia and also Christodoulou, following his suggestion, which was itself motivated by the interest of Jürgen Ehlers, then his mentor in Munich. Since then, Christodolou had worked on other problems. I was glad that another person other than me was applying his young energy to this one, under the efficient direction of a geographically close mathematician of undisputable value, Francis Cagnac. The resolution of this problem is complex, as all those concerning general relativity are. While Noutchegueme was working on his thesis, Cagnac and I discussed global problems for hyperbolic semi-linear equations. We proved and published in 1984 a theorem extending to a curved space-time, a property proved before, but only on Minkowski space-time. Before seeing our proof Leray had told me he assumed the inaccuracy of this result. An idea of F. Cagnac's had been essential for the completion of our proof. I used it again in a work in collaboration with Vince Moncrief. Thanks to a fellowship offered by the CNRS to my research team, Noutchegueme came to France for three months in 1990. We worked together on the coupled Yang-Mills-Vlasov system and obtained an existence theorem for solutions, local in time but global in the case of zero mass particles. After a brief interruption due to political difficulties in Cameroon, Norbert resumed mathematical research in Yaoundé and obtained, alone or in collaboration with a student, original results.

Another student of Cagnac, Marcel Dossa — originally from Benin but come to work in Yaoundé — worked, for his thesis and later, on the Cauchy problem with data on a charac-teristic cone for the Einstein equations. He obtained interesting results which remain classical for this important and difficult question.

During my last stay in Yaoundé, I told Francis Cagnac of my desire to visit, as a tourist, the famous Waza Park in the north of Cameroon. I had come to preside over the thesis committee of another of Cagnac's students, Hasannah, who had come from this region. Cagnac had me take a plane ticket for Garoua, the capital city of north Cameroon, which was an essentially Muslim region. Hasannah waited for me at the airport dressed in local fashion and looking very majestic in his long white robe, with a comfortable car borrowed with its driver from a relative who was an important man in the region. We stopped at this relative's place for a nice lunch which also involved Cameroonian personalities, happy, they said, to meet a French academician. Like Hasannah, this relative, too, spoke perfect French. No other woman was there. Hasannah told me it was not the custom. After lunch, we resumed our car trip to Waza. I don't know which offense had us stopped by a policeman. Hasannah, majestic in his long robe, got out. I don't know what he said but we then left. A quaint and cozy little house had been reserved for me in the park. The next day, a vehicle made for the visit of the park took me to see the animal residents who live there free and in peace: lions, giraffes and elephants. It was an extraordinary experience. I was, and am, full of gratitude for Hasannah. As I was about to thank him, he handed me a gift, a handbag in crocodile skin.

I did not return to Cameroon after President Ahidjo, an exploiter, they said, was replaced by Bija, who was considered at first to be an honest man and a savior. France stopped its cooperation, the CFA Franc was suppressed. Independent of these events, Francis Cagnac had reached retirement age. He came back to France without regretting leaving Cameroon, where the situation of professors had become as precarious as in Gorbatchev's Russia. It seems it is better now.

Benin (called Dahomey during the colonial period)

I made a short stay in Cotonou, the capital city of Benin, on the occasion of an invited conference to a congress of African mathematicians. The main memory I keep of it is the visit to the palace of the former kings of Benin. That building did not seem remarkable to me, but the explanations of the local guide during the visit were astonishing, especially when she praised the relative humanity of the last king who asked that at his death, his hundred concubines be killed before putting their bodies in his grave, while the previous kings had them buried alive.

China

My first travel to China was in 1982, to participate in a Marcel Grossmann meeting organized in Shanghai by Remo Ruffini.

There were then no direct flights to Shanghai. I stopped on the way, in Hong Kong, where I stayed for only one day. I went down to the sea to admire the view. I remained a little longer in 1987 in that city which boiled with activities and shops.

To organize his congress in Shanghai, Remo had managed to get the support of a remarkable man, Zhou Peiyuan. This Chinese physicist, born in 1903 to a family of Mandarins, had studied in the US and worked in various American universities, in particular on General Relativity in Princeton where Einstein was. Back in 1949 in China, when the Communist Party of China was in power, he was a professor, then the President of Beijing University. I don't know what happened to him during the Cultural Revolution. When I met him in 1982, Zhou Peiyuan was an important man in his country, Vice-chairman, I believe, of the Chinese central advisory committee. He was seventy-nine years old, in full possession of his intellectual faculties. He should have been retired but that was "not Zhou Pei Yuan", I was told. He was a very intelligent man. I had, with him, several conversations, always friendly, though I told him I did not share his beliefs on harmonic coordinates. Thanks, perhaps, to Zhou Peiyuan, the participants in this Marcel Grossmann congress were received in Shanghai in a way that scientists are not used to, among others things, by the mayor of this great city in a sumptuous banquet with all the refined Chinese specialties. Our Chinese colleagues had organized our transfer to Beijing. They kindly showed us around the Forbidden City and the picturesque old quarters, and took us to see the Great Wall, which had not yet been overrun by tourists. Very few cars in this great town were spoiling the visit; the great number of bicycles as well as the strictly conformist clothes of the inhabitants accentuated the picturesqueness. One felt the Party's discipline there. While I was walking alone in the streets, map in hand, a young man talked to me in English, offering to help me find my way. He said, "Lend me your map so it will be seen that I want to help you to find your way, not to make friends with you." During my last visit to Beijing in 1999, countless cars paralyzed traffic, and department stores sold jeans. I entered the shop where, ten years before, I had bought painted canvases representing the traditional "four seasons". There was nothing like that in 1999; the dealer assured me that such works had never existed.

It is on a large boat making the congressists visit the Shanghai harbor, that I met a couple of Chinese mathematicians, Gu Chaohao and his wife Hu Hesheng, who were essentially my contemporaries. Gu was a very high-level mathematician. He had done his graduate studies in USSR, as was usual at the time for the scientific elite. He managed, however, in English because before going back to China, he had stayed for a while at the Courant Institute in New York. There, he had proven the first global existence theorem for "wave maps". Hu Hesheng was an appreciated specialist of differential geometry. She had been Vice-president of the Chinese mathematical society. Unfortunately during the terrible Cultural Revolution, Gu and his wife were sent to work the land, separately, while their son was put in an institution for orphans. Gu and Hu did not like to speak about this period, which was, for them, so painful, physically and intellectually. Gu was not only a great mathematician, but a man of exceptional human value, respected by all. His perfect honesty, coupled with sound judgment, had given him an undisputed moral authority in that politically difficult period. Hu Hesheng was also a remarkable woman, direct and warm. I have been honored by their friendship. Our meetings in China and in France have been for me a pleasure and an intellectual and moral enrichment. They had the kindness to organize for me, on the occasion of my stays in Shanghai, conferences in other Chinese universities in cities where colleagues took care of my visits to interesting touristic places.

During my stay in Shanghai from September 12 to October 10, 1985, I gave a conference in Xian and was taken to see the famous archaeological site where is buried the emperor's terra cotta army. The city itself, with its large porticos, is worth the visit.

In Shanghai, I was lodged in a house for foreign visiting professors, near the Fudan University. The comfort was limited but the atmosphere was very nice. In 1987, I brought with me to China my daughter Geneviève, then a student. During our stay in Shanghai, we visited, accompanied by a student of Gu, the beautiful monuments of Hangchow and Souchow. At the end of our stay, Gu himself and his wife accompanied us in the descent of the magnificent gorges of Guilin. Geneviève and I went then to Canton and visited its picturesque market before going to Hong Kong to take our plane for Paris. We took the time to admire the bay and walk in the commercial quarters of this buzzing city. Geneviève was a bit disappointed not to be able to take advantage of it as much as she would have liked. For example, buying an ultra-modern bike and sending it to France was too complicated, so we did not do it.

Invited again in 1992, I expressed my wish to visit Tibet. There was no university there where Gu Chaohao could have me invited. However he organized, as well as he could, my trip to Lhassa. He had me invited to give a talk at Chengdu University, and took care of booking, for me, travel from Chengdu, a stay and a local guide in Lhassa, for a very reasonable number of dollars. I could thus realize what had been the dream of my mother and mine since reading a book of Alexandra David Neel: to go to Tibet. The guide was waiting for me at the airport and led me to the hotel, recommending me not to go out before the next day, to get used to the altitude. I disobeyed and went out alone that day and in the few following days. I walked in Lhassa, equipped with a map. I found the town miserable. I went to the Potala, following a path with a magnificent view. Inside, there was a queue of Tibetans passing before a monk, himself standing in front of an imposing statue of Buddha, in gold, I think. They were receiving a kind of blessing after some ritual gesture. Having arrived before the monk, I did not make the gesture and did not receive the blessing, but got an understanding smile which reminded me of the former Father Brillet. I have not found another source of spirituality in Lhassa in spite of my walks in the streets. I should probably have attempted to visit monasteries isolated in the mountains.

In 1999, I visited, again, Fudan University. Having to make a stop in Beijing, the Academician I was now wrote to the French embassy, asking for their help in the reservation of a hotel and transportation from the airport that, it seemed, had become difficult. An envoy of the Embassy was waiting for me at the airport and drove me to a hotel located in the hills near Beijing which was, apparently, the only one with available rooms. The hotel was not luxurious, but very pleasant. I had, the next day, a beautiful walk in the neighborhood and took a bus to go back to the airport and reach Shanghai. One must accept the bad, but also the good.

During one of my stays in Shanghai, I went to Nanjing for a few days to visit a specialist of differential geometry who had worked in France and was happy to practice his excellent French. I keep a pleasant memory of this town with facilities of human scale, that is, the very opposite of a megalopolis. Another time I gave a talk in Tianjin, which was organized by a friend of Richard Kerner's and supporter of France, Ge Mollin.

I happened to be there at the same time as Henri Cartan; he was accompanied by his wife Nicole, who had been greatly appreciated by her mother-in-law and my own mother. We were lodged in the same hotel, and took our meals together. Our Chinese hosts took us together to visit a zoo that housed pandas. I have — after many others — felt for Nicole Cartan, a devoted spouse who, nevertheless, did not compromise her values, esteem nor sympathy. She was walking alone in the city. It gave me the courage to do the same, and I took a bus to reach the city. I, unfortunately, have no sense of direction, so that after walking in the streets I did not know how to find the bus stop to return to the university. I know nothing of Chinese and passers-by did not understand English (nor of course French). One of them however understood my embarrassment and, no words exchanged, led me straight to the sought stop. Before going back to France, I stayed for a few days at a university in Beijing and I walked again in this city full of treasures of the past; it had changed a lot since my first visit in 1982.

My Chinese friends, Gu Chaohao and Hu Hesheng, invited me for a new stay in China, from 10 to 30 November 2003. I was already an old lady, still active, however. In Shanghai, I was no longer accommodated in a house with limited comfort for visiting foreign professors, but in a comfortable hotel near the university, frequented by Chinese personalities of the business world. During the last few days, I had the favor of a room in a hotel for tourists of the city center, to allow me to visit, more easily, its picturesque small streets. I had the pleasure of visiting my friends in the small apartment they had bought, thanks, they told me, to the salaries they received from the French university. These few months of salaries in francs would certainly no longer be sufficient now to buy an apartment, even a small one, in the center of Shanghai.

Gu Chaohao organized, for me, another visit at Nankai University, in Tianjin, this time upon the invitation of Chern who had settled there permanently. That exceptional mathematician, an American of Chinese origin and the recipient of several major awards, had come with his wife to spend his last years in the comfortable house he had built on campus. He was the object of all the care the Chinese know how to give to old people they worship. Chern was physically diminished, but his mind was clear enough to function in his day-to-day life. He was revered by all, even if his mathematical genius had lost its luster. He did me the honor and pleasure of taking me in his home during my stay in Nankai, and of giving me a dinner where we recalled the pleasant memory of our visit together of Strasbourg fifty years before.

The exceptional qualities of Gu Chaohao had earned him a nomination for President of the university of the great city of Hefei to which the government wanted to give the importance in agreement with its central situation. Gu Chaohao, a man of duty, had accepted this mission, and had completed it as well as could be expected. In 2005, he went back to Shanghai and, to thank him, important ceremonies were organized in his honor in Hefei. Gu took advantage of credits allocated on this occasion to extend another invitation to me in China. Afraid that my advanced age of eighty-two years would dissuade me from this trip, he proposed to have the travel and lodging expenses of my daughter accompanying me be supported in addition to mine. Geneviève could have asked for nothing better, but she in fact worked at that time for the bio-Mérieux society at a Franco-Chinese cooperation, and had to go to China anyway. Therefore, Geneviève did not use the travel expenses proposed by Gu Chaohao. She joined me in Shanghai after my return from Hefei and we went back separately to France. At the beginning of this stay, I met a great and friendly American mathematician,

John Ball — invited by Gu Chaohao and his former student, Li Tatsien, who had worked in Paris with my colleague and friend, Ciarlet. Li Tatsien speaks perfect French. He has created, in Fudan, a Franco-Chinese Institute. I listened to an interesting conference by John Ball and shared, with him, Li and Gu, an agreeable dinner offered by this Institute in a great hotel of Shanghai. During this dinner, John Ball told me that I had been chosen to give the "Emmy Noether" lecture at the next conference held by the international congress of mathematicians, intended, as the previous few, to encourage women to do mathematical research. Li Tatsien suggested to Gu to take advantage of the trip to Hefei to allow me to visit the Yellow Mountains.

I took the train for Hefei in the company of the Gu couple. I would have been quite unable to manage it by myself. They themselves had sent a secretary to take care of the tickets in Shanghai station, which was invaded by a crowd a hundred times worse than the one filling the La Pardieu station in Lyon now. The travel was comfortable, and the stay and ceremonies were very well-organized. They asked me to make a speech to an assembly of high-school girls to encourage them to do mathematics. I was provided with a translator to repeat, in Chinese, what I said in French. I don't know what she said, but I was warmly applauded.

After the ceremonies, Gu Chaohao and Hu Hesheng had organized a return by car with a guide, passing through the Yellow Mountains. The landscape of these mountains is splendid, and the trip across the countryside of China, where I had seen only cities before, was very interesting for me. Back in Shanghai, I found Geneviève; she had there a friend, a Chinese biologist colleague. With her, we visited Shanghai and neighborhoods as I could not have done alone. The development of this great city appeared to me to be fabulous. The majesty of the English buildings along the bundt has been preserved; the skyscrapers built on the other shore are beautiful. In 2005, the development of Shanghai seemed, to me, fabulous.

South Africa, Johannesburg and Cape Town

I made interesting trips to South Africa in 1996 and 1998 thanks to my friend Nail Ibragimov, a mathematician of great value and multiple interests, among which included differential geometry and non-linear partial differential equations. He was also an undisputed specialist of Lie groups' theory. Ibragimov — like many Russian scientists who could — had left USSR; the Gorbatchev perestroïka was a catastrophe for the scientists of the Soviet Union. The new regime reduced their salaries to almost nothing. The mathematicians of value have almost all emigrated.

Nail Ibragimov found a position as professor in the university of the greatest city of South Africa, Johannesburg. I accepted, with pleasure, an invitation to the colloquium he organized there in 1996. The plane I took made a stop in Cairo. Travelers then went on foot to the room where we waited to leave. I remember this short walk in the tepid and starry Egyptian night that made me regret not being able to linger for some time. On arrival in Johannesburg, a car was waiting for me and drove me directly to the university campus: a large field surrounded by high fences, with a gate guarded by armed soldiers. Behind these walls were university premises and comfortable lodgings for visitors. We were told never to go out of the campus, even with several of us together, without being accompanied by an armed guard. Therefore, I don't have much to say about Johannesburg.

Before leaving for South Africa, I had informed George Ellis I was coming. George is a relativist known worldwide, and a specialist of cosmology: a long time friend. I had met him years before in various summer schools and congresses; we had sympathized with each other. It was then the apartheid and George, a citizen of South Africa whose ancestors were not Boer but English, was opposed to this barbaric regime. He had done a thesis in Cambridge, England, under the direction of Dennis Sciama and had stayed there as a researcher or professor for several years. He had already become famous in 1973 for his collaboration with Stephen Hawking on the first book devoted to global structures of Einsteinian space-times. When the South African regime changed, George returned to his native country. In 1996, he was a professor in the Cape Town University. While teaching and continuing his work on cosmology, he took active care in helping in the modernization of the education system in his country. In spite of his busy life, George answered me quickly, inviting me to give a conference and organizing my stay in Cape Town. This stay was, for me, interesting and pleasant. I was told that in the daytime I could, contrary to the case in Johannesburg, walk alone in Cape Town. It was unwise only at night, but it is the same in some quarters of Paris. I took pleasure in the authorization, admiring the beautiful surrounding nature, thanks to public transportation. I gave a well-received conference at the university and had lunch with George who was as friendly as he had been in the past, though obviously occupied by important and varied tasks. On Sunday, he asked a friend to take care of showing me around. This very nice lady did it with good grace, making me penetrate the apparently relaxed and happy atmosphere where she lived in this beautiful area. I had ventured to ask her if she was not worried for the future, she answered firmly; no. I would have liked to keep contact with the nice lady with whom I had spent an interesting and pleasant day, but she did not wish it.

South Africa, Mmabatho and Pretoria

I don't know how, but Nail succeeded in leaving Johannesburg where everyday life for his wife and daughters was excessively difficult. He obtained a position in a smaller and more peaceful city, Mmabatho, and he had the kindness to invite me there, though I told him he had better invite someone who could be useful to him, as it was impossible for me because of my retirement. I keep a pleasant memory of Mmabatho — I went on foot from my lodging to the university on a country road where I met men and women, most of them smiling. After a few days in Mmabatho, I flew to Pretoria, a large city where order reigned, and from there, to Algiers, where I had never been. The atmosphere there was heavy and the looks of people in the streets, hostile. I would have been afraid if a young boy, almost a child, had not come to me, smiling and offering his help to visit the town. I accepted willingly and feeling quite secure in his company, I saw beautiful places in Algiers.

In fact, I had the pleasure of rewarding Nail Ibragimov for his kindness towards me. When organizing, in Karlskrona, Sweden, an international colloquium to honor Jean Leray, the organizer, Maurice de Gosson, had invited Olga Oleinik, but her state of health prevented her from accepting. Maurice wrote to me asking a suggestion for the name of a Russian replacement. I thought that Ibragimov, who had translated, into Russian, the Princeton papers of Leray, was a good choice. De Gosson willingly accepted my suggestion. He appreciated Ibragimov and, sympathizing with the difficult situation he and his family were in in South Africa, obtained, for him, a position in the technologico-scientific Institute Blekinge where he himself was working. Nail was happy to come back to Europe where he was appreciated by his new employers. His wife and daughters learned Swedish and adapted very well to life in Sweden.

Nail remained an active mathematician and organizer. He created a society working on group analysis and planned associated meetings. I had the pleasure of being invited to a meeting he organized in Nordfjordeid to celebrate an anniversary of the birth, there, of the famous Norwegian mathematician, Sophus Lie. I was greeted in Oslo by the professor Arnfinn Laudal, a great mathematician as well as an interesting and kind man. He listened with apparent and flattering interest to the lecture I gave at his university. Thanks to him, I visited interesting places in Oslo and could enjoy the splendid Norwegian countryside during the car trip between Oslo and the meeting place. I later went back to Oslo by myself. I stopped in Bergen and took a boat excursion along the beautiful fjord. I slept, for a moderate price, in the house of nice local people.

The relaxation of the Communist regime allowed Nail Ibragimov to divide his time between his native Russia and Sweden where his daughters settled and he still has an important position.

Brazil, Mario Novello

Mario Novello, my junior of twenty years, is a researcher specialist of General Relativity. He did important works in cosmology; his name is particularly associated with the "bouncing models". I met him in my first congress in India in 1975. Mario was then a dynamic and enthusiastic young researcher in the CBPF (Brazilian Center of Physical Research) in Rio de Janeiro, speaking French perfectly, with many varied interests in philosophy as in physics. We sympathized with each other. A few years later, Mario became Director of the CBPF and infused, in it, his vitality.

He invited me to come for a month to give, together with other guests, lessons on General Relativity. It was very interesting.

Rio de Janeiro is a very beautiful city with beautiful beaches along the Atlantic. Mario and his wife, whom I met there, treated me warmly, and I had an excellent stay. Invited again for a congress in 1984, I took with me my daughter Geneviève. I bought two "Brazilian air passes" that allowed us to visit, without ruining me, Salvador di Bahia, Recife then Belem where we met with the Amazon in a river trip in the vicinity. The most spectacular highlight of our trip was the visit, made at the insistence of Geneviève, to the Iguacu falls, an unforgettable experience in an enchanting setting.

Mario came, in his turn, to spend a month in Paris, invited by Paris VI University. We worked in collaboration on a subject that had interested us both: the obtaining of a regular conformal system for the Einstein equations, without using, as had been done by Helmut Friedrich, the Penrose spinorial formalism. Our result was published in a Comptes Rendus paper and detailed as a problem of my book, "Analysis; Manifolds and Physics II".

I went another time to spend a few days in CBPF, staying in a hotel near Copacabana beach. I look back with pleasure at these days and the magnificent setting of Rio de Janeiro that can be contemplated from the foot of the giant Christ dominating the city.

Venezuela

In 1982, I was chosen to preside over the Society "General Relativity and Gravitation". I hesitated to accept because I am little gifted for such functions. The previous President, Peter Bergman, and the General Secretary, Alan Held, insisted that I accept. Alan invited me along with Peter to his university in

Bern, for Peter's advice which enlightened and reassured me about my future duties. I was happy to see, again, forty-five years later, the picturesque Swiss city, and Alan efficiently reassured me on the consequences of my passage in the presidency. A pleasant consequence of this passage was Aragone's invitation to the congress of relativists from Latin America, in Caracas. For my conference in this congress, I resumed works of Eardley and Moncrief on global solutions of the Yang-Mills equations. For the acts, I wrote up a rigorous and detailed variant of the proofs. The accommodations of congressists and location of their conferences were in a pleasant suburb near the ocean; it was very nice and Aragone was quite friendly. He took us for lunch in a restaurant near the seashore, discouraging us from bathing and making it a "seaside restaurant", for sharks! However I did not see much of Venezuela; not even of Caracas and, to my regret, I lost contact with Aragone.

After Caracas, going up north, I wished to stop in Austin for the celebration of the sixtieth birthday of Bryce DeWitt and I did. On the way I stopped for two or three days in Mexico; my only visit in that country, though it is home to many valuable relativists. My best memory of Mexico is the visit of the museum. I did not see Yucatan, so beautiful it is said to be. I would regret it if regrets were not contrary to my philosophy.

Argentina

The triennial congress of the "General Relativity and Gravitation" Society took place in 1992 in Argentina, at Cordoba. The organizer, Oscar Reula, a mathematician author of interesting and rigorous works on Einstein equations, has always been kind to me. I wished, though I had retired from the university, to attend the congress. Some time before, I had received a letter from the third world Academy established in Trieste, which told me my inscription on the list of people and that it would cover the travel expenses for a conference in a third world country I was invited to. I think I owed this inscription to Jacob Palis, the famous Brazilian mathematician I had met in Coimbra; he was a member of that Academy. I knew an Argentinian physico-mathematician, Mario Castagnino who had been a member of our research team and was now an eminent professor in the Buenos Aires University. I wrote him and he invited me to give a lecture in Buenos Aires after the congress in Cordoba. He took the necessary steps for the payment of my travel expenses by the third world Academy. Oscar showed happiness of my coming and organized my stay, first a brief one in Cordoba itself, then in the mountainous place where relativists were meeting. The landscape was beautiful; unfortunately I was already too old to follow the other participants on their mountain walks. However, after the conference in Buenos Aires University which was general enough to interest non-specialists in General Relativity, I took a few days of freedom to visit that large and beautiful capital of Argentina, "good winds" in Spanish. I much liked this city, which seemed to me to be very European.

Korea

I had read a pretty book on Korea, "In the land of the calm morning." Gustave had appreciated his trip to Korea during his travel to the far east, organized by the Ministry of Foreign Affairs. I was therefore happy when I received an invitation from a Korean mathematical physicist, Yoon Hyuk Jong, professor at Konkuk University in Seoul, for a stay in his university. Yoon was responsible for a year-long program on General Relativity and the corresponding funds; he allowed some relativists to benefit from it, at a time of their choice. I happened to be in Seoul a few days after the arrival of Alan Rendall. I appreciate him as a scientist and as a human. I followed his advice to explore, using the subway, the surface and underground scenes of this great metropolis. Both are impressive in size and intensity. Seoul is a very large city, more modern than any other I visited, because, unfortunately for Koreans it was entirely destroyed during the war and had to be rebuilt completely. During the intersection of both our stays, Yoon took Alan and I to visit, in the neighborhood, a park where some buildings that recall the life of the past have been reconstructed. It was a beautiful and pleasant day.

New Zealand

An important professor in Cambridge (England), the astrophysicist Sciama, had sent me the work, which concerned questions I knew, of one of his students, Peter Waylen. He wished me to check the exactness of the results and give my opinion on their interest. I did the required work and gave positive answers in both cases, indicating a few minor improvements. Waylen was from New Zealand; his work made the content of a doctoral thesis and earned him a professorship in Christchurch University, located in his birth country. To my pleasant surprise, I received, later, in 1978, an invitation to give a plenary conference in the meeting of the mathematicians from the Australasia association. They were responsible for my living expenses and had arranged the payment of my travel costs by the French Ministry of Foreign Affairs. A friend and academician colleague, Jean Claude Pecker, who had benefited previously from an analogous situation, advised me to take advantage of my mission to go around the world, arriving at New Zealand from the east and returning by the West. I followed, willingly, his advice. Air transports permitted, and even imposed at the time, stops while not implying increases in fares. When heading to New Zealand, I stopped first in Singapore, then in Sydney, before reaching Christchurch.

In Singapore, I did not sleep in the famous hotel Raffles, whose rates had seemed, to me, excessive, but in a more modest and very pleasant one. I spent there, three well-filled days, visiting the city and its zoological garden full of charm. I took the cable car which, after a magnificent view on the bay, takes you to a small island with a pretty wooded garden. I also treated myself to a day trip to get to know Singapore's neighbor Malaysia. It was, at the time, akin to being in another world, its advancements and customs late by several centuries, but not without charm.

From Singapore I flew to Sydney, where I stayed only a few hours, just enough to visit the premises once used by the convicts deported from England. I then left for Christchurch, the capital city of the New Zealand south island. It was winter in that hemisphere. The bad weather, and what I had been told of that island — "It looks like Switzerland" — had discouraged me from booking a tour there. I keep, nonetheless, a good memory of the congress — the talks were interesting and the colleagues nice. Christchurch was a rather pleasant city; I remember a garden where, to my surprise given the climate, palm trees grew.

I had received an invitation to give a lecture at the University of Auckland, on the Pacific seashore in the north island where the former student of Leray, Philippe Dionne, was professor. He told me, "We have invited men, passing foreign mathematicians, why not a woman?" I took an interesting bus trip from Wellington, at the south end of the north island, to Auckland. The atmosphere was quite relaxed. The bus stopped for a while in a small city for us to drink the usual tea of our British friends. It stopped in the evening to let us dine and sleep in a small hotel at the edge of a picturesque volcanic lake. The two islands forming New Zealand are very different, as described in the novel of Elizabeth Goudge "The Green Dolphin Country". Unlike the first, the second is of volcanic origin; it was populated centuries ago by Maori, a people of warriors. I made an interesting stay in Auckland, thanks to Philippe Dionne, who took me to admire the bay and visit museums rich with memories of the island's history and stuffed remains of birds astonishing in their size; unfortunately they have now all disappeared because they were appetizing and not very combative.

Following my personal project, the world tour, I left Auckland for Noumea, renouncing the Fiji islands though they were on my way. I much liked New Caledonia, with its pleasant climate and magnificent scenery that I explored by bus, and Noumea, a city which I found in 1978 to be full of old-fashioned charm and peace. From Noumea, I went to Tahiti. I have not kept any memory of Papeete, only of beautiful landscapes in the area visited by bus. I took a boat to spend a few days in the nearby island of Moorea; I had rented a hut on the ocean front. It was beautiful, but I must confess that, as my friend Vince Moncrief told me he had experienced, I got bored rather quickly.

From Tahiti, I flew to San Francisco, where I was meant to find a flight to Paris. My plane was late and I missed the planned flight. It was Air France's fault, the company was therefore in charge of finding, for me, another solution. I was waiting at the counter when another passenger appeared. The attendant got busy taking care of him and turned back to me only when he had finished. He apologized for this gross injustice by saying, "If you knew how much he pays!", then to make amends he found me a convenient flight for the next day and a comfortable hotel room, the costs to be borne by the company, of course. So I finished my world tour.

Australia

My spouse, Gustave Choquet, had come back delighted from a three months' stay in Australia. I was, therefore, very glad when I received, in 1988, an invitation for a month at Canberra University where a renowned specialist of the subject, Robert Bartnik, was organizing a program on mathematical problems posed by the Einstein equations. I was told Canberra was a city without charm, but this assertion — which proved false — did not dissuade me to accept the invitation. However, I took an "Australian air-pass" that gave me free circulation on this continent-island. After a brief stop in Singapore, which had lost some of the charm it had ten years before, I landed in Darwin, a city located amidst superb tropical nature at the edge of the ocean. After two or three days, I flew towards the interior. I renounced Ayers Rock, which would have complicated my travels too much. Reading about that too-touristy place, does not alight my regrets. I stopped in Alice Springs, a town also located in the middle of the desert. I made an exotic and spectacular excursion in a land-rover with some other tourists. I admired the ability of our driver to find his way in this beautiful desert with neither traced road nor indication. I took some beautiful photos which still adorn my bedroom walls. I took the plane again for Cairns, where I rented a ride in a glass-bottomed boat, hoping to be able to admire the sight of coral, deemed to garnish seabeds in this region. Unfortunately the sea was rough and I did not see much of its depths. I comforted myself by watching a film in the hotel. I don't remember if it is before going to Canberra, or after leaving it, that I visited Brisbane, but I liked that town, where the weather was quite pleasant. I visited, with interest the world exposition held there during my stay, but I also took pleasure in simply walking its peaceful streets which had pavilions in nice gardens.

My month-long stay in Canberra is one of my best memories. The University buildings, including the one where we were staying in as well as the one where we had offices, were in a park full of blooming mimosas. I shared a well-lit office with my friend, Vince Moncrief, whom I have always appreciated for his intelligent and discreet company. The work atmosphere in Australia is not tense as it often is in the United States — scientific research is what it should be, a cooperative pleasure, not a competition.

When leaving Canberra at the end of our work program, I went to Perth where, organized by the Australian David Blair, one of the Marcel Grossmann congresses sponsored by Remo Ruffini was held. I made, on the way, a brief stop in Sydney, then in Adelaide, which I had been told was charming. I saw there, nothing remarkable, and I left, without regret, to go to my congress in Perth. I found there, with pleasure, many people I knew, among them John Archibald Wheeler and his wife Janet, and my friend Cécile DeWitt. With her, I took a boat to visit the renowned Rat island before going back to Paris. I did not visit Tasmania, which Gustave had much liked. I thought I should join my family.

Japan

During the Second World War, Japan was allied with the Nazis and its soldiers committed horrors in China; my mother and I read about them in the books of the great novelist Pearl Buck. I had, in the fifties, no sympathy for Japanese, nor my wish to go to Japan. This antipathy has faded with time, as the one I had for Germany did. I understood that one should not hold the progeny responsible for the horrors committed by soldiers of their parents' time. We do not condemn today's French for Palatinate's devastation by Louis XIV's soldiers. My mother had liked, in the sixties, her one-month's stay in Tokyo in the company of her son François visiting there for work. They had stayed in the Franco-Japanese house, then directed by the mathematician Delsarte, who was the former dean of the Nancy University where François was teaching. My mother had kept a very good memory of her stay in Japan. Hence, I accepted, with pleasure, Remo Ruffini's invitation to the Marcel Grossmann congress he had organized in Kyoto in 1991. I particularly appreciated, as a tourist, the brief stay I made in Nara in Japanese housing, visiting temples, which for me seemed to be of a very exotic beauty. After , I spent a few days in Tokyo, housed in a large room of the Franco-Japanese home, thanks to the kind director who was replacing Delsarte.

I returned to Japan in 2001, to a colloquium organized in Tokyo by the Franco-Italo-Japanese group working on partial differential equations: Vaillant, Gourdin, Colombini, Murthy, and Spagnolo. I had the privilege of a room, which was small but meant only for me. The colloquium was interesting, but the city of Tokyo did not leave me with much memory. After Tokyo, I spent a few days in Tsukuba, invited by a Japanese colleague I had met in Italy. He was a very friendly man and his work interested me. I wished to invite him to France. Unfortunately, he died prematurely. I did not return to Japan.

18

OUR HOUSE IN DAMMARTIN

Descendants of my maternal ancestors continued to live, for several generations, around Mantes-la-Jolie. I myself spent my youth and later, a good part of my holidays, in the large house belonging to my parents in Dammartin en Serve.

Dammartin en Serve

Dammartin en Serve is fourteen kilometers from Mantes-la-Jolie on the departmental road 11, at the boundary between two regions. One is with wooded hills and valleys where flow two small rivers, the Vaucouleurs and the ru d'Houville. In a valley lies the small town of Septeuil, 5 km from Dammartin, and in another, the village Montchauvet, located a little more than 1 km from Dammartin as the crow flies. Montchauvet was important once; there had been a castle of which there are only remnants, and a beautiful Roman church built in the XIIth century, which was partially destroyed in 1909 by a thunderstorm, then by a clumsy blasting to secure the place. It is still a beautiful monument, though its present square bell tower, the result of the action of Jean Richepin, who was then mayor of Montchauvet, dates only from 1912. Montchauvet is a popular tourist destination. A picturesque path descends from Dammartin to an old bridge called "pont de l'arche" (the arch's bridge) which allows walkers to cross the brook of Houville and to go up between shrubs and fields until they reach Montchauvet. On the other side of Dammartin, a national road penetrates into the Mantois plain, which borders a pond abounding with fish called "la mare aux Prévôts". It then borders a bunch of chestnut trees surrounding an ancient sculpture called "la croix aux sorts" (the cross of spells). The road continues across a plain covered with cultivated fields to the village of Longnes, which is 3 km from Dammartin; a small secondary road leads to the hamlet La Fortelle, 1.5 km farther. Another departmental road crosses Dammartin, between Boinvilliers and Fleins Neuve Eglise, then Tilly. I have been on all these roads a lot, whether on foot or bike.

My Grandmother's House

The house inherited by my grandmother from her parents is on the main street of Dammartin, in front of a place surrounded by lime trees where the church is located. This church is not very old and the statues which decorate it are not classified as historical masterpieces. In my youth the church was always open and had its own priest. My father accompanied us on Sundays to Mass.

Behind my grandmother's house, there is a small garden. She was very fond of it. She was proud of her pear tree which grew along the wall; its Williams pears were, for her, the best in the world. She liked the perfumed flowers of her lilacs and cultivated those on a narrow flower bed for as long as she had the strength. She loved fuchsias; I do, too in memory of her. In my childhood, a tub in a small hut was used as a toilet, and water came from a well in the village. The house had, on the ground floor, a living room with a window overlooking the street, behind a small and dark kitchen. A corridor led to the small garden and a staircase led upstairs. There was the master bedroom, with its window overlooking the street, and a small anteroom lit by French doors, where my uncle slept in his youth, when he came on holidays to Dammartin. I slept there sometimes, so that my grandmother, when she became very old, would not have to stay alone. At the mezzanine level, there was a room overlooking the garden. It was called the pigeon-house. As a child, my mother shared this room with her own grandmother. She had kept an excellent memory of this cohabitation; she loved her cheerful and communicative grandmother. Circumstances led me to sleep in the pigeon-house on some holidays. It was a small room, but quiet and pleasant.

Our House

When I was a young child, we spent a few summer days in my grandmother's house, but it was too small to accommodate us all. The neighboring house was a former abbey with a porch and a few ancient remains, among them a column and a beautiful stone table in the garden, which was surrounded by barns. That house was vast, but in a bad state — its roof threatened to collapse onto that of my grandmother. My father was ready to file a lawsuit to require repairs when the owner, more or less bankrupt, offered to sell it. My father, who liked Dammartin and to please his wife, accepted to buy this big house. It was done, and the necessary repairs were made by a mason, a very friendly local boy with the family name "Mauve" and first name "Guy", given by parents with humor (in French, guimauve means marshmallow). Guy Mauve had started as a baker, but lung problems had led him to change his profession for masonry, apparently a healthier business.

During our childhood, the family has made, each year, happy stays in this big house. We took the train at the Saint Lazare railway station to Mantes, then a bus to Dammartin. When Jeanne and I have been able to, we went with our father on bicycle from Mantes to Dammartin. The way there is a great descent, then a ride from village to village between meadows and forests. I liked this ride, Jeanne liked it less.

My grandmother still lived in her house, but came to share our meals. My sister and I were very happy that each of us had our own bedroom. Mine, called the blue room because of the color of its walls, overlooked the street surrounding the church. I had my table and a seat, a cupboard and a bed; I was home. I spent, there, many happy hours, reading or writing at length to my friends. A door opened to a few steps that mounted to a passage above the porch, then a few steps descended to the great bedroom of our parents, followed by a smaller room where François slept in his youth, and finally a staircase going down to a room stocked with books which was called "the gray room". It opened on one side to the garden, on the other side, to a great room with windows overlooking the street and garden; this was my father's office. Another door of my blue bedroom opened to a fairly large room where one arrives at a straight staircase going down to a vast room, which was the dining and living rooms, with windows to the garden and on the street, and at the extreme end, a fireplace, under which was an enjoyable wood fire we often assembled near when the evening was fresh. Opposite my bedroom was another room followed by a smaller room, both overlooking the garden. It was and remained, all her life, my sister's domain. After her marriage, she gave the smaller room to her spouse. As for me, who did not have my elder sibling's authority, I had to leave the blue bedroom after my father's death, as my mother found it more convenient than the conjugal room.

After her death, it was François who settled there. As for me, with or without husband, I occupied various bedrooms. Each had their advantages and disadvantages. Finally, I was given the master bedroom of my grandmother's house. It was OK. François had installed central heating in parts of the house and had it extended to that room to please me.

Childhood Holidays

My parents liked nature and walking. Each afternoon when the weather was nice, a family outing was almost mandatory, sometimes enhanced with a particular snack. A frequent destination, only a few kilometers away, was the pretty little path which goes from Dammartin to Montchauvet, passing over the "pont de l'arche". In my childhood there was, in Monchauvet, a small café where we stopped and had the right to enjoy a dried sausage sandwich; it was a feast! We also walked to Septeuil, and a little farther to the inn, "La Mare Aux Clercs" (The Pond of Clerics), where the traditional snack was hot chocolate and brioche, much appreciated by us walkers. Many other lovely walks without gustatory bait were available, in the woods and hills which surrounded Dammartin. In my childhood, there was also access to quarries full of fossils, which proved fascinating for young collectors.

When our parents had other occupations, I played with my brother or sister, or the two of them: throwing a ball above the porch, playing croquet in the garden or organizing, on the few steps and the small terrace in front of the living room, snail races or a make-believe farm, with chestnuts for cows, rosehip fruits for pigs and sloes for poultry. When we were a little older, my father bought a ping pong table. He put it in the largest of the barns; called "the hangar". I often played ping pong with my father.

In the evening, when my father was not at work, it was often Rami at the family table.

Visits to Cousins

Each summer, we went on foot to Longnes and la Fortelle to see our farmer cousins — that is, three families. Two of them had a farm in La Fortelle. The third was a woman who lived alone in Longnes, Valentine. Her house was well-kept, small but not tiny, with a flower garden in front. She asked for news of "Aunt Louise", my grandmother, who did not accompany us, too old for the required 7 km on foot. Valentine gave, to my mother who was about ten years younger, cooking recipes the latter listened to with feigned interest.

Visits to other cousins were more interesting, to the greatest of the two farms. There lived three generations. The seniors, Anatole and Marie, in their sixties, owners of the place, were very friendly to me. Both were intelligent and rich in anecdotes — especially Anatole — and complacent listeners — especially Marie — who interrogated me on my life in Paris while introducing me to milking cows and raising chickens. She awoke, in me, a desire for the vocation of farmer that life did not lead me to practice. I don't regret it because, as it is said of many things, "it is not the same now." Anatole and Marie had an only daughter, Agathe, a robust woman of about forty, who was open and welcoming too. Her husband, Georges, was less communicative. It is true that the heavy work he assumed, including plowing by leading the ox, was energy-consuming. They had an only son, Denis, a fat — in fact, obese — boy a little older than me, intelligent like his grandfather and an excellent student at the village school. His teacher advised him, as formerly had been done for my grandfather, to continue his studies and become a teacher, but that profession had lost its prestige. Moreover, during the war, farmers had, in a sense, been kings, having the precious resources necessary to life: food. We benefited from it during the scarcity in the years of the Second World War. My father took, every week, the train for Bréval, made by bike the 8 km ride to La Fortelle and the 8 km ride back, with some food to improve our scanty meals. I did it a few times myself. Often, during holidays, I went from Dammartin by bike to my cousins' farm to seek food difficult to find elsewhere. These trips were not a chore for me but pleasant excursions. I liked, moderately, the bowl of milk my cousin Marie offered to "this poor under-nourished Parisian", but more than that, I liked pedaling in the country and chatting with my cousins. For one reason or another, my cousin Denis remained, therefore, a farmer in La Fortelle. He married and had two children, a girl and a boy. The life of those, however, became more and more difficult, like for other small farmers, and both eventually abandoned the profession. Denis died young, my own life became complicated, and I lost sight of that branch of my family. I was recently told by my niece Françoise, who now owns the family house, that after some interruption, this farm in La Fortelle is occupied again by descendants of my cousins who rear rabbits and poultry "bio" (organic) which they sell on local markets. I am glad of that.

I don't know what became of the other farm, also in La Fortelle. We went on the same days to see these cousins, represented by the second generation, Solange and her husband whose name I forgot, and their only son Gérard, who was several years younger than me. Solange was a woman, young still and full of energy. She raised sheep. They became attached to her by affection. Solange told us that her sheep had gone on a hunger strike when she had left for a few days. Of the husband of Solange, I remember only that he talked often of events that happened and customs which were observed during his military service, that he had performed in Ponchartrain. He said, "When I was in Ponchartrain" as we would say, "During our stay in China". I was very surprised when, some years later, going by car to Dammartin, I went through Ponchartrain, about twenty kilometers from La Fortelle.

After the War

Our house has been occupied during the war, first, by a group of soldiers who said they were Austrians and did not show hostility towards us when we went there one summer. However, an occupation unit later made much damage, among others, burning the furniture, perhaps to be warm in winter.

After the war, my mother obtained, from the competent organization, subsidies to repair the damages to the house, and the necessary work was done. In the refurnished house, the three Bruhat children and their mother enjoyed family meetings with their own children and even later, with the oldest of their grandchildren. My brother François — who remained a bachelor, a Parisian and was passionate in gardening — became the administrator and main user of the house, with his mother for as long as she could, then, alone. The three Bruhat children and their families continued to meet regularly in Dammartin, in particular, traditionally, during the All Saints' vacations. My grandson Raphael remembers these holidays and the very meaty meals prepared by his great-uncle François.

Separation

I don't remember exactly when François convinced us — my sister and I — to sell, like him, our shares of the house in Dammartin, keeping the usufruct, to our niece Françoise who lives near Paris, so that the house remains in the family after our deaths. I accepted his request, mostly to please him. I thought then that, much younger than his sisters, he would survive us, which was not the case. After selling my share, for less than its value, I believe, I continued to spend some comforting weekends in Dammartin, wandering on foot or bicycle in its familiar surroundings, for as long as I could drive my car, because there is no more bus from Mantes to Dammartin.

One of the last times I went to Dammartin, driven by my daughter Geneviève, was for the funeral of François, sad and moving, where the family gathered. Two of my grand-nephews played, in the church of Dammartin, appropriate music. A priest of foreign origin made a commonplace speech. As many French parishes do, Dammartin now shares its priest with neighboring parishes; its church is not very old and the statues which decorate it are not classified as historical masterpieces. However, this church is only open on the Sundays during its turn to host the celebration of Mass, because sacred objects which decorate it have become tempting to thieves.

I think now of Dammartin with some nostalgia, but, as was said by a writer, I don't remember which, "A happy memory is often on Earth truer than happiness."

EPILOGUE

One can see in my biography that I had had a full life, three children, seven books, some three hundred articles and many friends. I have lived my life from day to day, obeying responsibilities to my family or my work and my own feelings that I have not always been able to dominate.

My three children are, I believe, all attached to their mother but very busy — Daniel, by his research in biology and his work as Director of a big Institute; Geneviève, by her job as a doctor, always working to expand her skills, and also by her artwork as a painter and a sculptor. Michelle is a little more free, though very busy with her role as practising and fighting ecologist. They are responsible adults, each a father or mother of three children. I leave it to them to write their own biography if they wish, especially because experience has shown me that memories of the same events kept by differents actors or spectators are sometimes very different. It is said that it is our mind which builds the world we live in; there is truth in this statement.

I will limit myself here to a few generalities. I generally wished well to people I met. My mother recalled that Christ has said "love your neighbor as yourself", and that the neighbor is the one next to you; she added "and who bothers you." I have not fought for great causes, humanitarian or others.

I was not obsessed by ambition, nor feeling that I had something to prove, perhaps because I was a woman and, at least in my youth, the weaker sex should retain this term not to shock. However, I have been happy with my successes, especially because of the warm friendships that accompanied them.

The play is over, I wait only the last act, death — that is, my diving into the unknown. But I use the first person pronoun without knowing its meaning. The assembly of these billions of cells in my brain where memories accumulated which constitute the actual me will not continue to exist in a prolongation of what was, for me, time until now. Present science tells us that what is for us, reality, is only an image at scales of space and time, of an extraordinary, more complex reality. As was said by Thibault Damour in some interviews, science is the human activity most able to give precise information on the universe we live in, but he adds that he has the deep feeling that the ultimate bricks are still, and will long remain, obscure. Also, science sometimes says how, but not why.

The ascertainment, beautifully expressed in this thought of Blaise Pascal, is still true.

"I do not know who put me in nor what is the world nor myself, I am in a terrible ignorance of all things, I do not know what is my body, nor my senses, nor my soul, and even this part of me that thinks what I say, considers everything and itself, does not know more. I see those frightful spaces of the universe that surround me and I find myself attached to a point of this vast expanse, without knowing why I am in this place rather than in another and why this little time that is given to me to live is assigned in this point rather than in another of the whole eternity before me and the whole that follows me. I see only infinities on all sides that surround me as an atom and as a shadow that last only a moment without return. All I know is that I must soon die, but what I know least is this very death which I cannot avoid".

Blaise Pascal

To conclude less solemnly, I will quote the prayer advised to me by Father Brillet during the brief period when he was my spiritual director, "My God, if you exist, save my soul if I have one." I did not practise it much, not knowing the meaning of the words "God" and "soul". To finish relaxing my readers, I shall tell them the wish told to my five-year-old brother after his appendix operation by a canon hospitalized in the same clinic. "Life is short, but it is long; life is sad, but it is merry. There are two kinds of soups, the lean soup and the fat soup, it is the grace that I wish you." (In french, fat is "grasse", which is pronounced the same as "grâce" of "the grace of god".

Acknowledgements

I thank my daughter, Michelle Fourès and my admirable friend, Thibault Damour for reading and correcting the French version of my book.

I thank, for checking the English translation, Vincent Moncrief (prologue and epilogue, chapters 16 and 17) and Jim Brown (the remainder of the book).

Printed in the United States
By Bookmasters

Printed in the United States
By Bookmasters